Unrealized Digital Democracy

Unrealized Digital Democracy

A Critical Analysis of Power in the Digital Age

Garrett Pierman

LEXINGTON BOOKS
Lanham • Boulder • New York • London

Published by Lexington Books
An imprint of The Rowman & Littlefield Publishing Group, Inc.
4501 Forbes Boulevard, Suite 200, Lanham, Maryland 20706
www.rowman.com

86-90 Paul Street, London EC2A 4NE

British Library Cataloguing in Publication Information Available

Library of Congress Cataloging-in-Publication Data

Names: Pierman, Garrett, 1993- author.
Title: Unrealized digital democracy : a critical analysis of power in the digital age / Garrett Pierman.
Description: Lanham, Maryland : Lexington Books, [2023]. | Includes bibliographical references and index.
Identifiers: LCCN 2023034471 (print) | LCCN 2023034472 (ebook) | ISBN 9781666935578 (cloth : acid-free paper) | ISBN 9781666935585 (epub)
Subjects: LCSH: Internet in public administration. | Democracy. | Information technology--Political aspects. | Political participation--Computer network resources. | Internet--Political aspects | Social media and society.
Classification: LCC JF1525.A8 P555 2023 (print) | LCC JF1525.A8 (ebook) | DDC 351.0285/4678--dc23/eng/20230809
LC record available at https://lccn.loc.gov/2023034471
LC ebook record available at https://lccn.loc.gov/2023034472

∞™ The paper used in this publication meets the minimum requirements of American National Standard for Information Sciences—Permanence of Paper for Printed Library Materials, ANSI/NISO Z39.48-1992.

Contents

Acknowledgments

First, I would like to thank the Graduate School at Florida International University. The generous funding of much of the research for this work, through their Dissertation Year Fellowship, was instrumental to the creation of this work in its early form. These thanks should also be extended to the Green School of International and Public Affairs at Florida International University, which has been my intellectual home since 2011.

Second, I would like to thank the members of my dissertation committee, namely Clement Fatovic, for their invaluable feedback through and beyond the dissertation process. Much of that feedback has sharpened this version of the work substantially and will guide my research beyond this work.

Finally, I could not have completed the work without the support of my friends and family. In particular, Antonella Avogadro has been a stalwart partner through this (very!) long process. Nicolas Ortiz and Ian Baker have provided much-needed friendship through the sometimes uncertain and lonely journeys through our early careers. My parents were, similarly, more than patient in listening to the struggles of their first-generation doctoral student in the years-long process of writing a dissertation and then a book. I dedicate this work, ultimately, to my late grandmother, Marjorie "Ruth" Pierman. Your commitment to critique, and to standing up for one's ideals no matter what runs through this work to its core.

Introduction

THE NEW NORMAL

On Wednesday, January 6, 2021, I distinctly recall sitting at my desk with a sprawling Word document open that would, eventually, become my dissertation. Given the impending departmental deadlines, the plan for the day was to sit at my desk and write. For several years prior to that Wednesday in January, I presented a series of term papers, followed by conference papers, and, eventually, defended a dissertation proposal that promised a history of the internet, influenced by a set of new-materialist political thinkers, that forwarded a more contingent view of technology that commonly is held by many folks: that the internet and technology more broadly improved human lives through communication and that the empowering potential of a digital, democratic public far outweighed the risk of privacy becoming an outmoded concept.

Nearly everyone I shared early versions of that work with found it interesting: by 2021, the fact that digital tools had been at the fore of politics was clear, as events ranging from the Arab Spring in 2010 to the then-most-recent American presidential elections proved that, for better or worse, the internet was now a central element in political life on a global scale. Furthermore, many of my colleagues, friends, family, and much of the human species had from about March of 2020 onward, become mostly digital creatures who worked, played, loved, and lost through their screens. The argument that I made in those early works was a simple one: the internet can be best understood as a set of actors who interact with and shape one another through relationships that are marked by power dynamics. Further, the balance of power on the internet has changed over time. Where once the state (specifically, the United States government) had reigned supreme over an early version of the internet, corporations now reigned supreme and not only use the internet to pursue the ends of profit but in so doing, shape the behaviors of the state as well as ordinary people (even those who do not use the internet). This corporate-led reshaping of the state and people through digital tools strongly resembles, I find, the kinds of feudalization that Jurgen Habermas documented in his widely lauded work *The Structural Transformation of the Public Sphere*.

What I saw on January 6, 2021, deeply troubled me. Americans, many of whom were deeply unhappy with the outcome of a free and fair election, attempted to storm the Capitol of the United States of America. It was the first organized act of violence at that particular location in two centuries. In a year already marked by much fear and uncertainty, a sense of exhausted, terrified paralysis overcame many of us witnessing what many assumed was an attempted coup by an angry mob. What on Earth, I could not help but wonder, would cause so many of my fellow Americans to believe the clear falsehoods peddled by Donald Trump, their chosen candidate, so thoroughly that they committed politically motivated violence against the government in an attempt to install Trump as president despite his clear electoral loss?

To formalize the puzzle that the January 6 insurrection brought to my mind, I attempt here to answer the following question: is there something structural about the internet (through which the aforementioned lies spread) that causes otherwise peaceful, if not reasonable, people to commit political violence? If there is, then developing an understanding of those structures and how they came about will prove invaluable to creating more peaceful democratic political processes going forward. The events of January 6 were a stress test for democracy in the digital age: we must develop an understanding of the fault lines in the foundations of our democracy, I fear, to forestall broader political violence in the future.

To address those questions, I offer in this book a mapping of digital politics as it is experienced in the United States in the second decade of the twenty-first century. To offer such a mapping, I first develop an analysis that serves as a framework for understanding political events such as those absurd and awful events of January 6 as an effect of feudalized relationships between actors that make up a network. I posit that contemporary digital politics can be well understood as a series of power-laden relationships and that those relationships affect the composition and behavior of all the actors within the digital actor-network. In exploring the asymmetries of those relationships as they have evolved over time, I explain the processes that led some of my fellow Americans to take up violence against our shared government.

The remainder of this introductory chapter makes the case for an analysis of digital politics that draws from the work of Bruno Latour's innovative actor-network theory and furthers a new materialist understanding of political relationships. This mapping of digital power not only relates actors in geographic relation to one another, as Latour does in most of his work but also takes into account power dynamics that affect actors' ability to think and make political decisions in ways that affect political behaviors. Each of the subsequent actors follows one of the relevant sets of actors: the state, corporations, algorithms, and internet users. In each of these chapters, I relate the actors to one another in a critically informed history of digital political life that makes the case that, by January 6, 2021, many internet users experienced the internet in such a way that it is hardly surprising that they believe the unbelievable and act upon it, all the while corporations make massive profits with the sometimes unpredictable aid of algorithms, and that the state that once sponsored the creation of the early internet now feels, quite literally, the wrath of what it wrought and ceded to corporate interests.

THE INTERNET AS AN ACTOR-NETWORK

Slovenian political thinker Slavoj Zizek remarks, critically, the then-new film titled *The Matrix* highlights a major problem of the internet: rather than a "Big Other" against which we can unite and struggle, the internet exposes us to a nearly incomprehensible universe of "little others," leading us to struggle, fight, and argue with one another rather than against any common Other, as folks in the West might have found first in fascists, then communists in the twentieth century (Zizek 1999). These small struggles, he predicted in 1999, would likely to lead to a proliferation of conspiracy theories of increasing absurdity to help individuals and small, tribalistic groups make sense of the incoherent world in which they now find themselves (ibid.). This runs somewhat contrary to the path chosen by some of the more democratically minded public policy thinkers who have been, for some time, speculating as to whether or not the web can be used to the ends of practicing democracy (Simon et al. 2002). Astra Taylor, in her work noting that the web as it is now has become dominated by corporate interests, laments the lack of a more critical history of the web as well (Taylor 2014, 3–7). Similarly, Cass Sunstein finds a relative paucity of histories of the web that explain in detail how, exactly, we have ended up with echo chambers full of divisive rhetoric rather than an inclusive digital agora (Sunstein 2007). There has similarly been, Cristian Vaccari finds, a lack of scholarship or public discourse that focuses on how users and nonhuman actors can, and do, influence each other on a global digital scale; most works now are American-centric and privilege corporate narratives, preferring stories of business success and increasing shareholder profit (Vaccari 2013). These few varied examples serve to draw a broader point that I wish to make here: social science and philosophy have not yet come to any strong conclusions about how people will use the internet to relate to one another, even several decades after the internet became publicly available in the early 1990s. This paucity of a consensus among social scientists adds to the concern I expressed in the opening pages of this chapter: there is a distinct need for a reasonable explanation for the decidedly undemocratic character of some of the politics that seems to come from the internet today if those of us who would see democracy survive can work toward a better future.

The use of computer networks such as the internet, and the data generated on those networks, need contextualization and critique if we are to better understand the normative implications of specific technologies (Boyd et al. 2012). Thus, this chapter advocates for a new understanding of the internet in terms of Bruno Latour's actor- network theory, modified with the innovations of new materialist scholarship as understood in the perspectives of Gilles Deleuze, Manuel DeLanda, and others hold that social, physical, and ideological entity as contingent results of processes of production that occur within the bounds of the material world at a particular place and time. It is my aim that this analysis, which I term actor-network-influenced analysis, uncovers structural relationships in the makeup of the contemporary internet that can help explain the processes by which people might be led to commit acts of political violence when those acts could have otherwise been avoided.

The thinking of Bruno Latour is enlightening in exploring the agency of material in the formation of assemblages. His actor-network theory (ANT) bolsters the topographical project of new materialism by providing a set of patterns and trends for which to look in the historical construction of actor-networks. Latour's development of ANT is mainly descriptive but lacks some explanatory power in terms of describing asymmetry between actors in each actor-network. To explain the apparent asymmetries in digital power that have implications for the establishment of deliberative publics online, I deploy an adapted version of ANT that makes room for asymmetries: this is in keeping with Latour's thinking in some of his clearer explanations of ANT, as well as secondary scholarship that aims to bring ANT into conversations with materialist thinking.

Having made the case for conceiving of the internet as an actor-network, I propose, in subsequent chapters, an understanding of the web as an actor-network in which the state, corporations, internet users, and algorithms all play a vital role in creating contemporary political conditions.

DEVELOPING AN ANT-INFLUENCED UNDERSTANDING OF THE INTERNET

Given that the internet can be seen as a set of relationships in which goods and services are produced and consumed, with nearly incomprehensible profits being generated along the way, one set of methods that appeals to many scholars such as Astra Taylor, Shoshana Zuboff, and others is historical materialism. After all, historical materialism came about, in its original formulations by Karl Marx and Friedrich Engels as a method by which to explain the emergence of new forms of economic and political life that rests on the assertion that those forms of life are composed of both material objects and the humans that relate to one another through those material objects (Marx and Engels 1972). Much of the political theory concerning the internet directly engages in class-based analyses from a variety of methodological and ideological perspectives. For example, Astra Taylor's illuminating work, *The People's Platform*, argues that the development of the internet led to a rearrangement of the ways in which labor and consumption are done but does not represent any sort of revolution in terms of ownership of the means of production and the accompanying concentration of political power (Taylor 2014). Other analyses of digital class focus on the hope that tech giants will improve our lives out of benevolence (Jarvis 2011) or observe them as subjects of biographical interest rather than as members of a class with interests in profit and power (Smith 2012). Work that seeks to explain the origins, structures, and current practices of the internet and digital life more broadly should be able to take seriously both the material objects that make up the network as well as the ideology, ideas, and other nonphysical elements of that same network. Historical materialism of various kinds, then, highlights the internet as a set of relationships, mostly focusing on the relationships between people (as consumers) and corporations (as owners of platforms or other means of production of physical and digital goods and services). Vitally, historical materialists of all sorts focus on disparities of power: from the beginning of Marx's understanding of history as a series of class struggles to more contemporary accounts such as that offered by Taylor, historical materialists find much of their well-founded concerns with the disparities in the power of actors to affect outcomes within a given political system. This insistence on an analytical eye on power dynamics is a valuable tool I borrow from the historical materialist tradition. Where historical materialism, or its offshoot, Marxism, might leave something to be desired in terms of the potential of nonhuman actors, new materialism works toward this in thoughtful directions.

As a body of thought, William Connolly describes new materialism as analyses that

> encourage us to identify shifting elements of ontological uncertainty and real, conditioned creativity in the periodic intersections between several forces in the world. There is often an ebb and flow in most domains, as a system goes through a period of relative equilibrium followed by another of radical disequilibrium. A philosophy of becoming set on several tiers of temporality does not, though some fools project such a conclusion into it, postulate a world in which everything is always in radical flux. (Connolly 2013, 401)

This places new materialism not as a single school of thought, but as a general epistemological approach that recognizes that the objects of philosophical and political inquiry are in constant motion and to understand them an understanding of the processes by which that change occurs is critical. New materialism, in its focus on the "periodic intersections between several forces in the world" does not present a total break from previous materialist thinking: that intersection, because it is "creative" is still dialectical in the Marxist sense. Where new materialism finds new ground to explore, I concur with Connolly, is in its mappings of periods of (dis)equilibrium in dialectical terms rather than looking for overarching narratives of thesis, antithesis, and synthesis that can be found in Marxist thought, especially that written by Marx and Engels. Innovating on the insistence on anthropocentrism that many new materialists find within fields ranging from political science to biology, new materialists instead attempt "to problematize the anthropocentric and constructivist orientations of most twentieth-century theory in a way that encourages closer attention to the sciences by the humanities" (Gamble et al. 2019).

In the context of this work, the stakes of a new-materialist analysis demand an examination of the construction of the internet as a set of material objects, created in a social context by actors, but also influence that social context—specifically in terms of the (dis)empowerment of human subjects that are both grounded (in their identities, communities, nationalities, for example) but are also, simultaneously, in flux in part because of their interactions through the internet. In so doing, I contend as Braidotti expresses as central to a new materialist understanding of phenomena to develop an analysis that follows, "concrete yet complex materiality of bodies immersed in social relations of power" (Braidotti 2012). People experience this materiality, in Braidotti's analysis, as a nomad subject, who is not only moving through material reality but is changed by it, and further changes that environment: some of this motion, I argue, conditions actors' relationships to and through the internet.

Turning more directly to a new materialist conception, Manuel DeLanda founds his understanding of materialism upon a rejection of essentialism:

> But if one rejects essentialism then there is no choice to answer the question [of the identity of material objects] like this: all objective entities are products of historical projects, that is, their identity is synthesized or produced as part of a cosmological, geological, biological, or social history. (DeLanda 2012)

With DeLanda's assertion in mind, I focus on the consolidation of the internet with properties of its own with a keen eye on properties that (dis)empower persons, as nomadic subjects, to participate in democratic deliberation through the consolidated, but still in-flux, internet. As Wendy Hui Kyong Chun develops in *Programmed Visions*, the software with which users interact affects the agency of those users in terms of connections, memory, and perception (Chun 2011, 77): the goal of this project is to map the relationships that describe the effects of those changes on some internet users, leading them, for instance, to commit political violence.

Both historical materialism and new materialism focus on power-laden relationships and ground those relationships in material relationships between people or groups of people. Commonly, the older materialists speak in terms of class relationships (Marx's classical formulation of the struggle between the bourgeoise and proletariat, for instance), and new materialists allow for, in Connolly's estimation, periods of both equilibrium and flux rather than revolutionary brakes. Those relationships and flows of power are, I find, vital to understanding the internet as we experience it today as users, scholars, and professionals. What materialisms thus far lack is a deep understanding of the technologies that make up the internet as political actors in their own right. Here, actor-network theory stands to enrich materialist analyses in ways that give greater explanatory power than either materialism or ANT can offer on their own as political worldviews.

ACTOR-NETWORK THEORY AS CRITICAL METHODOLOGY

Though new materialists do follow some general trends, which I have noted thus far, it is an eclectic, contested territory of intellectual exploration rather than a cohesive school of thought. Thus, instead of claiming to present a history of the internet from *the* new materialist perspective, I instead outline *a* new materialist perspective, which I use as guidance for the following history of both the internet and the human subjects which have co-construed one another over the past several decades. Specifically, Bruno Latour's presentation of what he terms actor-network theory (ANT) informs a nuanced analysis of the processes by which agents co-create one another, which speaks well to the new-materialist perspectives of Gilles Deleuze, Felix Guattari, Manuel DeLanda, and Jane Bennett (Bennett 2010). ANT, quite deliberately, borrows from literary theory in the mandate to tell a story as accurately and sensibly as possible (Latour 2000). Thus, when literature, here broadly and generously defined, offers the challenge to build narratives that take the material and the ideological seriously as foundations to narratives and explanations, theory would do well to respond by adding serious consideration of material as co-constitutive of political phenomena.

The materiality of the web, from a new materialist perspective, is central to the productive processes that create not only the web but the human subjects that relate to the internet. Latour holds that human beings intervene in productive, material processes at frequent intervals to construct a world of facts, norms, and objects (Latour 1986). Keeping the human at the center of analyses of processes that appear objective, such as the publication of scientific papers, ANT insists that humans and nonhuman objects are both equally necessary to the construction of what Latour refers to as an actor-network (ibid., 11–14). Meaningful experiences are had in the interaction between people and material objects which, for both Latour and the new materialists so far surveyed, co-constitute that experience in an iterative fashion that has been, is, and cannot help but be open to change. To illustrate this point, allow me a brief example of a series of events.

Examine the two following sentences:

1. Donald sends an angry message through Twitter.
2. Twitter spreads Donald's angry message to his followers.

In the first sentence, the actor, Donald, is the doer of the deed. As a human being, that he can affect changes in the world around him is apparent in the ability of Donald to tweet. In the second, however, Twitter, a nonhuman thing (or, more accurately, a network of humans and nonhumans), is doing the spreading. These examples show two of the core assertions of ANT. First, it does not come across as immediately foolish for a nonhuman actor to act as an agent that can effect change in the world (ibid., 71). The second is that human and nonhuman actors often work in the same networks (ibid.). Taken together, these assertions undergird Latour's analytical stance in ANT: human and nonhuman exist together, and both participate in the shared construction of networks. The human and nonhuman participants in the network depend on one another to act. If we remove the human actor, Donald, then his phone merely resides on a desk, and no angry message spreads through Twitter. Removing the human from the network, in this example, renders the network itself inert. Similarly, were there no phone to type with, then the message could not reach Twitter. Without Twitter, even if a message existed on a phone screen, Donald would not be able to share information in the same way. To have Donald able to tweet requires both the presence of the human actor, Donald, as well as Twitter, and all the electronic devices that connect Donald, Twitter, and his followers.

This small example demonstrates one of the central points of ANT: both humans and nonhumans act as agents in the construction and experiencing of networks. Latour's ANT, he notes in *Laboratory Life*, is an anthropological lens through which to study the sciences, but since fruitfully applied to many realms of science and technology studies, it is not a strict set of methodological rules. Instead, it forwards an epistemology that privileges neither the materialism of so-called objective reality nor the conventional construction of intersubjective reality found in some poststructuralists, such as Baudrillard (Latour 2005). Both, Latour develops over the body of his work, are co-constitutive elements of our lived, social realities. That epistemology, which has us examine the complex relationships between types of data instead of insisting on beating reality into neat categories and doing untold damage to the narratives that we end up constructing in the process, insists on taking both the materiality of our subjects of study, as well as the interventions and intentions of human actors equally seriously (ibid., 64). Speaking to the tradition of critical theory, from which Latour hails, ANT attempts to keep both empiricism and the tools of theoretical critique well in hand to better develop a picture of the world that places human beings as mediators acting in a material world, albeit themselves mediated by the constraints of that materiality as well as their ideologies and psychologies. Through ANT, one can understand phenomena in terms of a cast of actors, both human and nonhuman, which exist in relationship to one another in the material world (ibid., 159).

Note that actors need not necessarily be human beings. Examine the Twitter example again: the digital platform of Twitter (itself an assemblage in new materialist terms), which is not human, is as central to the expressing of a digital message to an audience as the human actor who composed the tweet. Put differently: Twitter's existence, passing a message through content filters, and dissemination of a tweet makes it as much an agent as Donald is in the eventual sharing of the message. To make the example even clearer, we can remove human beings entirely from the immediate series of sentences and have a program write the message for us, which will then be tweeted out just as if Donald did it himself. Journalist Andrew McGill has, in cooperation with *The Atlantic*, developed a project that assesses Donald Trump's tweets. A bot seems to write some of the tweets, usually to make announcements at odd hours of the night (McGill 2017).

Latour explains this well in one of the first works of ANT, *Laboratory Science*, where he makes the case that science is not simply human scientists discovering preexisting facts about their objects of study. Instead, he finds, human scientists construct facts in a social setting, with complex goals, and in a network of actors that include the scientists themselves, a broader academic/professional community, and the matter that the scientists study. In claiming that the matter being studied, the broader academic/professional community, and the scientists themselves all co-create the findings of research presented as facts, Latour seeks to both decenter anthropocentric narratives of the discovery of facts and call for richer narratives, including a larger cast of actors in the creation of any given piece of information, object, or scientific result (Latour 1986). Yes, the material of the stated subject matter is one such actor, but there are others as well that affect the scientific process strongly.

The production of scientific facts, for Latour, entails much more than the relationship of the scientist to the object of discovery. First, for Latour, scientists depend upon a mutually intelligible cultural context: this shared context of known and accepted ideas (the boundaries of which science seeks to expand through their so-called discoveries) is what designates scientific activity as such and, using Latour's example, differentiates scientific activities from food preparation or the reading of entrails for prophecy reading (ibid., 42–44). Latour's insistence on the shared context as a precondition for both doing and understanding science on behalf of readers and the observers deployed in the research of scientists for *Laboratory Science* has an important methodological impact on the history of the internet I conduct here.

This analysis opens a methodological path that I follow in the rest of this work. For Latour, the material elements of a laboratory (the raw materials, animal subjects, tools, etc.) allow people to produce facts in the form of papers. It is then the role of the scientist to convince and be convinced or, in other terms, to make meaning and share it with others. This process is iterative and done over time within the contestation over the creation, acceptance, spread, citation, and validity of facts. Thus, Latour suggests a method of research: the anthropologist observing the lab produces a book detailing the processes by which laboratories produce facts as actor-networks. I follow a similar track in the subsequent chapters that maps how people, corporations, the state, and algorithms, in creating and using digital platforms, (dis)empower one another in terms of their ability to participate in democratic political life.

From the perspective of ANT, one can ask questions about how a fact or material object came to be: the fact or object that human actors create goes on to influence the relationships between human actors in the future. This analysis differs from older, Marxist materialism in the emphasis on the agency of nonhuman actors: Latour, in developing ANT, challenges some of the more anthropocentric notions of Marxism's focus on classes made up of human actors. The processes of the production and sharing of those facts and material objects, when considered in the preexisting social contexts, illuminate the power-laden relationships between sets of actors. Actors, both human and nonhuman, condition one another through what Latour terms "programs of action." These programs of action influence what actors relating to one another can do and what effort might be necessary for an actor to take certain actions. For instance, as Latour develops, a speed limit sign, a speed bump, and a police officer all affect a driver's ability to speed, with differing effects and consequences (Latour 1992). Programs of action, I argue, are also built into digital platforms that internet users encounter online. Additionally, if one holds that material objects and facts that influence human beings are themselves given meaning in the context of continually changing actor-networks. In other words, in mapping the creation and implementation of the internet within larger social contexts from a new materialist perspective, I aim to evaluate the trajectory of that development and implementation in terms of the (dis)empowerment of varying sets of actors: delineating the relationship between the actors is where we can begin to map digital power.

Latour's conception of what counts as an actor, here, is also of vital importance to the development of a critical history of the internet that takes matter seriously. Keeping in mind his commentary on Plato, which maintains that material objects are "indifferent to our quarrels, our ignorances, and the limits of our representations and fictions" (Latour 2004, 13–14) a brief example might bring into greater clarity a new materialist conception of matter as it relates to (dis)empowerment of human beings.

Take, for example, a hammer. One human being can, of course, make a hammer initially to build a home. That same hammer, either in the hands of that person or another person, facilitates the destruction of some other material object. At worst, a murderer or member of a shadowy government agency may make use of the same hammer to inflict harm on other people, disempowering them from living normal lives thereafter. Different humans may and do use tools to differing ends. The existence of the hammer once it has been made, again returning to Latour's commentaries on Plato, inhabits the second house: it simply exists. To what ends the hammer will become deployed, much in the case of scientific facts, is up to the interactions of human actors with that object. One can, despite this unpredictability of the eventual usage of tools, learn a fair bit about the agents behind the tools by examining the stated or implicit intentions of those who first manipulated matter into a specific configuration. First, if we untether the tool from that agent that put purpose into it, we can learn the aims of the agents by studying the object, effectively retracing the steps used above to sketch purpose on a conceptual level. There is a wide variety of hammers that have wildly different programs of action, even if they consist of the same base materials. These programs of action, if we relax Latour's insistence that ANT considers all actors to be symmetrical (Castree 2002), give insights into the power dynamics of actor networks. Let's continue the example with a few types of hammers.

The first is a framing hammer. Typically, this is what comes to mind if one does not place a modifier in front of "hammer"; likely, you own one already. These hammers are of moderate weight and have two distinct features—a strike face and a claw for pulling nails. The purpose that an agent places into a framing hammer, I infer from the name of the thing (more on naming in a bit), is the framing of walls. In the object, then, we see that these two features correspond to that purpose. With the first, the strike face, careful observation would reveal that the face has a cross-hatched pattern on it. Aside from being great at worsening injuries, an unintended feature, this cross-hatching on the face keeps the head of a nail from sliding under the force of the blow, more effectively transferring the energy from the agent into the nail. Second, modern construction techniques for raising wood-framed walls eventually require the pulling of a great many nails: thus, a claw attached to the rear of the hammer ensures that the agent, in this case, a framer, has the right tool at hand without needing to carry or locate a second one: this increases the overall efficiency of the agent.

In examining both objects, the only real difference between them is the shape of the implement opposite a striking face. One could begin to make several inferences based merely on the presence of those objects, and the context in which they inhabit. If, for instance, one could find a hammer on a workbench at a job site, then it would be reasonable to infer that there are roofers and framers afoot who have these purposeful objects present to assist them in fulfilling the wishes of an agent—to eventually make a building. Removing those objects from that context and placing them in another, for instance, in a toolbox in a shop, changes the inference we can make about them then. We reasonably infer that in finding both of those objects in the same toolbox stored in a shop, we have found the shop of a general contractor, who may use either of those tools.

Mainly, I use the example of the hammers to make an analytical assertion that guides the rest of the chapter: if we can learn about the purpose of an object, and the intentions of the agents who made that object, from the examination of those objects, then we may begin to do so in the case of the contemporary internet as well. Thus, I assert that corporate actors, operating in the context of the internet, exhibit some of the effects that Latour attributes to capitalism as a set of affective foreclosures (Latour 2014): the massing of profit as the good is a choice corporations make and it conditions what programs of action these corporations come to build into online platforms. Users, then, are affected by these same programs of actions in ways I develop later in the work, which limit some users in terms of their ability to think critically and clearly. Once viewed as such, it becomes possible to begin to address the fitness of the systems of the web that we have now to address the needs of other agents. If we, as agents seeking to make democracy are handed or already have tools that have been designed to implement different programs of action, then we might find that those tools come up short in similar ways that roofing with a framing hammer is slower, more difficult, and often deeply frustrating as the agent continually bumps against the limitations of the tools at hand.

In a similar vein, the original intention of the creator, or set of creators, of an object, is sometimes divergent from those of later users of an object. Returning to Latour's example from "Where Are the Missing Masses," a driver may well wish to speed and will find varying degrees of both frustration and conditioning of their behavior by objects programmed by the state to stop speeding: a police officer cannot stop speeding but will raise frustration through ticketing, and a speed bump can physically stop speeding lest the driver risk damage to their car's suspension (Latour 1992). In mapping digital power accurately, then, the contestation between actors and their varying programs of actions is vital to a full understanding of the dynamics of those power-laden relationships.

Painting the picture of the internet will require, in keeping with the methodological attitudes of ANT's interpretation of new materialism, three things. First is an understanding of the nonhuman components of the network: in this case, the computers and connections that make up the physical parts of the network. Establishing a shared, if brief, history of those nonhuman actors in the network is the goal of the remainder of this chapter. Second, in the following section of this chapter, as well as the rest of the work, I place human actors (users, owners of websites, and others affected by digital practices) in their relationship to the nonhuman actors and structures of the web and each other. Then, third, I can begin to examine the relationship between the human and nonhuman elements of this network to draw a larger picture and begin to address more directly the fundamental question that drives the work as a whole: how these human and nonhuman actors are configured in such a way that makes digital democracy so seemingly difficult when compared to the hopes held by thinkers in the decades preceding our own, which I codify in terms of the normative benchmarks presented in the preceding chapter.

Materialist explanations of capitalism offer rich descriptions of effects that impact the actors that make up the contemporary internet, especially corporations that own social media platforms. These capitalist effects, namely the concentration of capital, dependence on labor for the generation of profit, and resistance to constraint by the state affect how corporations behave in the contemporary internet; they are the programs of action that the corporate actors design digital platforms to follow. A version of ANT that slightly relaxes some of the assumptions of Latour, namely the symmetry between actors in terms of power (Soderberg and Netzen 2010), recognizes that these capitalist effects exhibited by corporate actors asymmetrically affect concentrations and flows of power online. What Latour's ANT lacks, I find in concert with Johan Soderberg and Adam Netzen, is in accounting for asymmetries between actors. In the case of the internet, for instance, describing an individual user and Meta as merely co-constitutive would miss out on the clear asymmetries in power to influence their shared actor-network.

Instead, I suggest borrowing materialist conceptions of power, which do focus on asymmetries and exploitations between actors, to adapt Latour's ANT to the study of digital power. Deploying this adapted version of ANT, however, allows me to provide an analysis that describes the ways in which some users experience social media platforms in ways that modify their agency substantially. I suggest, then, making the following adaptations to Latour's ANT to map digital power more accurately:

1. Recognize capitalistic effects on the agency of actors, especially corporations.
2. Emphasize asymmetry: some actors, especially those who enter networks before other actors, may have more opportunities to implement programs of action that come to affect other actors later.

These two modifications allow for actors to affect one another and for me to keep the contributions of materialist scholarship in mind when describing power dynamics. In so doing, I aim to map the topography of digital power in such a way that I can evaluate the internet normatively against a Habermasian ideal.

The internet is at least in part made up of nonhuman objects; it is in smart phones, computers, wires, data centers, and, increasingly, refrigerators and children's toys that we, the human actors, have decided to link together across the globe. It is also in less immediately physical nonhuman mediators—the software platforms, user interfaces, and code that condition the flow of data between physical objects as well as the human actors. Getting a clear idea of how these nonhuman mediators came to be is vital to the development of a full picture of the internet as an actor-network. Humans created these nonhuman actors to accomplish set goals: in understanding those goals and the eventual machines made to fulfill them, we will come to more carefully learn how these nonhuman mediators have come to affect human political life.

Considering the internet as an actor-network, I assert, gives analytical leverage in addressing the original question of this chapter. If the internet can be seen as a network in which people shape machines, and those machines further shape people, then there is likely an explanation in which the shaping of people by the internet as we see it today leads those people to a state where they are more likely to commit acts of political violence than they otherwise would have been.

OUTLINE FOR THE REMAINDER OF THIS WORK

The first chapter of this work pursues two main objectives. The first is to document the early history of the internet, from its inception as a Cold War–era project in the 1960s, to the 1990s when corporations successfully commercialized the network. It was the state that was the dominant actor during this period of the internet's history—corporations were mere contractors and most everyday people had no interaction with, for example, ARPANET, at all. This early internet, like highways, served as a nationalizing project of the Cold War era. This early version of the internet had little direct impact on the lives of ordinary people. This changed rapidly in the 1990s.

The second objective of the chapter is to follow corporations' rapid commercialization of the internet in the 1990s. While users on this early commercial version of the internet were mostly consumers, the dot-com crash of the late 1990s led, I claim, to a feudalized structure of the contemporary internet that casts most users as prosumers, folks who produce and consume information nearly constantly in a feudalistic relationship with corporations that might leave little room for agency and political action. Some of these prosumers' agencies became so affected that they attempted to storm the U.S. Capitol in January 2021.

Chapter 2 takes up this prosumerism as people experience it, through the lenses provided to us by Paul Virilio. Taking Virilio's work on speed and political life in the post–World War II era, I claim that prosumers on the feudalized internet are experiencing politics far too quickly for our brains to keep up. This outpacing of our psychology generates the possibility for political accidents such as an attempted coup at the behest of a former reality television show host.

Chapter 3 continues this argument in conversation with the work of Daniel Kahneman, whose work on the relationship between the intensity of thought and energy necessary for said thought finishes the absurd portrait presented in high resolution to most of the world in January of 2021—when people have no time to make sense of the world around them critically, they might well turn to the political violence that was suggested to them to keep them as loyal prosumers of 24/7 news. I ultimately conclude that it is one of the major, predictable consequences of a feudalized digital political that prosumers, if fed a steady diet of fear, rage, and alienation, will sometimes commit political violence that has no basis in rational thought. The violence I saw is the result of people who have been, in effect, socialized into being digital serfs to the extent that they mistake the content that they like, subscribe to, and share, for fact-based information to the extent that otherwise reasonable human beings are willing to attempt to overthrow their government based in conspiracies whose authors argue, in open court, are too silly to be taken seriously. Instead of an informational utopia in which we can learn, and understand one another better, a feudalized internet is a dystopia of our own making in which we are both making and suffering from not only a pandemic of respiratory illness but of feudalized politics that enriches corporations as it starves the critical thought necessary to democratic life with participating citizens.

Chapter One

A Critical Digital History

THE STATE IN THE EARLY INTERNET

In this chapter, I examine a period in the internet's history in which the power dynamics between the state and corporations changed dramatically; as the role of the state in what would eventually be called the internet waned, the relative power of corporations waxed. That shift in digital power dynamics, which occurred in the early 1990s, also saw the addition of ordinary people as internet users to the stage. While most people had no contact with the precursors to the internet regularly, the centralization of power that the state pursued in this period was adapted later by corporations in the commercial internet. The lessons these corporations learned, and then implemented, in the contemporary internet have had major effects on shaping internet users as prosumers ill-equipped to discern political truth from fiction.

The first actor whose story needs to be well mapped in making the case for the internet as a power-laden actor-network is the state. Specifically, the internet as we know it today is an evolution of a military network that the United States government created in the mid-twentieth century, and it did so at the same time as it created another network to similar ends. Those ends, as I make the case for in this chapter, establish the state as an actor that seeks to empower itself to the ends of defense, though the means by which the state does so can be co-opted to the ends of other actors. As the role of corporations online grew in the 1990s, the state took a relative back seat in digital power, which, I argue at the end of the chapter, allowed corporations to feudalize the internet into spaces where corporations have much sway and influence over the experiences of users.

The earliest version of the internet was called the Advanced Research Projects network, or ARPANET for short. ARPANET, in its original form, was funded by research grants from the Department of Defense; this, too, was the case with the Military Network, MILNET, and the National Science Foundation Network, NSFNET. As a string attached to that public funding, ARPA stipulated that contractors and grant recipients could not use the network for personal or commercial purposes (Moschovitis et al. 1999). This became something of an issue after the NSFNET and MILNET split in 1980, which left the former lacking funds. In 1991 the National Science Foundation concluded via an internal audit that, if scientists' work could go on unhindered, the commercial interests that had long cooperated in the development of the NSFNET could use it for other, profitable purposes. Thus, with the stroke of a regulatory pen, the NSF opened the internet up to commercial use and began awarding contracts to facilitate the development of private, subscription-based internet services on a national scale (the Internet Corporation for Assigned Names and Numbers 1991).

The process that led the NSF to that decision was, Jay Kesan and Rajiv Shah note, a largely opaque one in which companies that had been involved with federal contracts since the ARPANET influenced the process of privatizing the internet in ways that gave those corporate actors influence over not only governmental processes of privatization but the shape that the internet came to take in the 1990s (Kesan and Shah 2001). This change came after over a decade of waning funding for the NSFNET, while MILNET's funding from the government was along the same generous lines as the early ARPANET. To survive, then, NSFNET would need to expand possible means by which to gain income, and without the generous government funding they had enjoyed during the ARPANET era, the businesses that had codeveloped the computers and software of ARPANET and NSFNET were more than willing to step up as the major actors of a civilian internet by providing already-developed and newly developed services to individuals and companies now that the strings that often came with scientific research grant funding were cut. With networking services now legally allowed to use the same technologies that corporate actors and academics developed for ARPANET and its offspring, physically connecting these new home computers through phone lines and onto massive servers, the commercial internet had begun.

The commercial opening of the internet is the point at which it entered the broader public consciousness, but in terms of setting the stage for the internet as an actor-network, it is a point further along in the tale. That same nation-state that founded and funded early computer networks, in the mid to late twentieth century developed other networks simultaneously to the ends of centralizing and maximizing state power. The opening sections of this chapter chronicle the development from ARPANET into the early commercial internet to the end of documenting the power shift. On ARPANET, the state was the primary actor that made the decisions in the development and use of these technologies. After the internet's commercialization in the early 1990s, however, corporations took an increasingly dominant role online.

ON COMPUTERS, HIGHWAYS, AND THE STATE

By its commercial opening in the early 1990s, the network that would become the internet went by several names: ARPANET, MILNET, and NSFNET all refer to the iterations of the actor-network controlled by the United States government's various branches, with the cooperation of universities and private companies. Today, "the internet" is the most common name for the network built after NSFNET, though, before the acceptance of that term, others, such as the Information Superhighway and the World Wide Web were in more common usage.

So, for the internet, what is in a name? Two names for the internet give some clue to the features of the internet that we see now that are cause for some level of analysis here. Firstly, I take up the "information superhighway," a term popular before the turn of the twenty-first century. In calling the internet a superhighway, one involves the context and image of the other, physical highways that have states built on physical territory over the last century or so. The archetype for the modern highway was the German Autobahn. Examining this original, modern superhighway may well give us some clues as to the intent and understood context of the people who used that name to describe the internet. The programs of action that states attempt to implement in highways may well repeat in other actor-networks as well.

The Autobahn, a product of the German state beginning in the interwar period, was a project that, in several regards, exemplified the balance of power between the state and corporate actors in Germany at the time. It was at least in part a political project that was meant to unify the German state physically and to do so also ideologically, by connecting disparate places more closely and quickly. This attempt to unify the German nation-state around a large project that would make it physically easier to move from one place to another was fundamentally aspirational for a state trying to both maintain the recent unification of the state itself, as well as recover from the insurmountable financial ruin of the World War I and concurrent Great Depression. The building of the Autobahn was, in this sense, an aspirational metaphor for the goals of those who funded it, a German bourgeois class willing to invest in massive state projects even before the consolidation of Nazi power (Rollins 1995). Interestingly, this act of creation was simultaneously an act of destruction. In making this new superhighway, the state destroyed no small amount of the German countryside and remade the modernist image of the new Germany, with its consistent aesthetics in roads, signage, and service stations making manifest the ideological dreams of a unified state. This came also, as most road projects do, with the destruction of forests, natural habitat, and the world as it generally was (ibid.).

In the United States, a similar project was underway as well. The motivations for the expansion of the American highway system were multiple and echo those we will see again in a later exploration of the early days of the commercial internet in its shift from a government project. Early proponents of a federally supported highway system touted its military significance, economic benefits, and its probability for creating a powerful symbol for the arrival of the country as a modern, unified nation (Bement 1916). In the interwar period, Franklin Roosevelt took this wisdom to heart and was more than happy to use the construction of roads as a jobs program as part of the New Deal (Reading 1973). The commitment to the building of a modern interstate was redoubled through Eisenhower again, now as president, with the passage of the Federal Highway Act of 1956. The law committed the federal government to spend, in cooperation with states, what was necessary to modernize American highways (Griffith 1982).

As a long-running technical and bureaucratic endeavor, the Eisenhower administration brought local and state actors into the process of planning highways, promising funding mostly, with federal money (Rose and Mohl 2012). What this federal inclusion in highway planning did, in some instances, was provide room for actors outside of the government, with more local political interests to make use of the technology of highway construction to further ends of racializing the geography of Miami. In the case of Overtown, the building of a highway paved over a community quite literally and served to disempower its members.

Overtown, a community in Miami, Florida, went by the unofficial name of the Harlem of the South in the early twentieth century, and it lived up to that moniker in many ways. Home to mostly Black residents, many of whom worked in the service sector in Miami's more affluent neighborhoods, Overtown was, into the early 1950s, a center for Black life in Miami. This life was a vibrant one, consisting of homes, businesses, and cultural centers that reflected Miami's singular mix of influences ranging from Afro-Caribbean to the Deep South (Institute of Government at Florida International University 1998).

Sadly, one of the elements that Miami borrowed from some of its Deep Southern roots was racism. Overtown is some of the choicest real estate in the area in terms of geographic centrality in the greater Miami area. Just to its South are those affluent neighborhoods I mention above; to the East, Miami Beach; and to the west, Miami International Airport. If you want to get anywhere in Miami, or to Miami in the first place, your path will likely run straight through Overtown. For the majority-white county government, incorporated in 1956, to accommodate the needs of a growing Miami-Dade County, Overtown being at the center of all that new growth posed a problem that needed solving. As if on cue, the federal government, pockets flush with funding, came asking for local "expertise" and input for proposals to extend the interstate through Miami and down to the Keys further South.

The federal government took two proposals somewhat seriously. One had the new highway going through the Florida East Coast Railway Corridor, which would have put I-95 next to the water, just east of Overtown. The second was to pave over a thriving Black community that had been the target of highly racialized campaigns against proposals for public housing and other community development efforts. Instead of making use of a railroad track whose best days were long behind it, the state and federal governments agreed to run intestate routes right through Overtown. The project began in 1956, coinciding with the passage of the Federal Highway Act.

For residents of Overtown, the interstate meant something different, and much worse. First came the evictions, then the bulldozers. What was once a thriving community for people of color now existed as memories or in the shadows of a complex of overpasses. Under those overpasses, where homes, businesses, and lives once unfolded relatively normally, you will now find a homelessness crisis. As a tourist or a resident of the rest of Miami, you would be able to ignore this, though, as you drive right over Overtown, admiring the giant glass obelisks and billboards that make up the Miami skyline today, blissfully unaware of the history that you are driving over.

In the destruction of Overtown, we see an example of the development of an infrastructure project that provides a means by which the state creates a homogenous experience of flow, in this case of vehicle traffic, along the routes and by the rules set by the state. By setting the routes and the rules, a productive process, the state reifies a relationship of power, with that state as an agenda-setter and, in Overtown, a racist county government consulting. The new highway over Overtown occupied a moment of simultaneous production and destruction. That destruction, which defined the boundaries of the assemblage of the highway privileged capitalist development of Miami over the lives of poor and working-class Black folks. In 1950, 45 percent of the Black population of the county lived in Overtown. A decade later, Overtown's population, and rates of homeownership, dropped by half. Residential use of land in Overtown, by 1990, virtually ceased to exist (Institute of Government at Florida International University 1998).

This example demonstrates the ability of nonhumans to serve as agents of change in power dynamics within actor-networks. People drew the plans, but it is the road itself that paved over Overtown: the interstate now sits over what was once a thriving community. The original planners have likely since died of old age; they can no longer act as agents. The road, however, still affects lives above and below its surfaces: Overtown remains, as of this writing, a deeply impoverished community with sparkling examples of postmodern architecture casting long shadows on what was once a thriving center for Black life in Miami. In terms of the analysis I developed earlier in the chapter, the Overtown example demonstrates that corporate interests have, historically, used technologies developed by the state to affect the agency of others. Specifically, the corporate program of action, oriented at profit disempowered the people living in Overtown from continuing to live their lives as normal.

These major projects, whether highways or ARPANET, demonstrate that the state, during the twentieth century, was more than willing to work with corporations to the twin ends of enhancing the capabilities of the state as well as making private profits possible. Where the highways, especially in the case of Overtown, quite literally paved over the lives of ordinary people, ARPANET was still something of a far-off Cold War project from the perspective of the average American. Something would need to change to result in the prosumer-filled actor-network that I observe on the contemporary internet.

THE EARLY COMMERCIAL INTERNET AS CONTESTED TERRITORY

Highways were but one context in which, during the twentieth century, the contestations in power between the state, corporate interests, and ordinary people plate out. In the opening section of this chapter, I followed the history of the predecessors of the internet, the ARPANET, MILNET, and NSFNET, as the state created them to the ends of national defense and forwarding science. Along the way, corporate actors not only helped to develop the networking technologies but made use of state resources in the development of those networks and others to begin forwarding their ends as well. Within the context of the mid to late twentieth-century development of those networks, the state took an early role in using those networks to reify and solidify state power, but, as NSFNET commercialized in the early 1990s, the stage was set for a shift in the balance of digital power that could potentially empower corporate actors through the now-commercial internet: Because corporations had an earlier role in the internet than people outside of the relevant corporations, universities, and government agencies, those corporations began to set programs of action into digital platforms that continue to affect users today. These affects, I develop in the chapters that follow, cast people as prosumers who are ill-equipped to discern misinformation from facts, while simultaneously taking that misinformation to be vitally important and requiring immediate, sometimes violent action.

With the relative retreat of the state from the commercial internet, there, too, was an opportunity for ordinary people, users who bought or accessed internet subscriptions through computer terminals enabled with graphical user interfaces, to participate in digital life as well. The introduction of the commercially available internet made room for some fascinating human stories that exemplify broader trends of digital power.

This section takes up a critical history of the internet from its commercialization in 1991 until roughly the present. In walking through that two-decade period, I make the case that on the early commercialized internet, corporate actors tried out new strategies to expand their influence, power, and profits on the internet.

Taking up one of those human stories, focusing on the period just after the opening of the internet for commercial use, highlights some of the dynamics that have come to influence contemporary online political life. Josh Harris was part of the first generation to grow up concurrently with the internet. When he was born in 1960, ARPANET had not yet been created (Smith 2012). Harris focused his early career on the technology of his day, and he became a moderately successful communications consultant for the radio and television broadcasting industry. Things got a little more interesting in 1993 when Harris founded Pseudo, an internet video and radio broadcasting company on the newly commercialized internet. By the end of the decade, Harris was throwing lavish office parties, engaging in postmodern performance art pieces, and attracting large sums of investment money from venture capitalists, all to the optimistic cheers of media outlets (Bunn 1998). So far, the story was not unusual for the mid- to late-1990s; companies registering sites with the domain suffix "com" were popping up, and being sold, and their often young and inexperienced executives were making fast millions. And they celebrated those millions with gusto. Socially, this period added to, and complicated, the myth of the American entrepreneur, with Harris and others like him working long hours, taking large financial risks, and, in some cases, becoming, at least temporarily wealthy.

Interestingly, these early dot-com businesses often paid their staff in stock options; this allowed these businesses to operate on shoestring budgets, but it made things especially risky for workers. Just under half of tech industry employees in New York in 1999, for example, were paid at least in part in stock options, and many worked under "nonstandard" arrangements such as being considered contractors, freelancers, or part-time workers. These arrangements left workers paying for much of their training and taking on a great deal of economic risk to participate on the internet in its early commercialized period (Neff 2012). The potential upside was that, were the dot-com enterprises sold either privately or publicly, those previously worthless stock options could be sold for real cash. Before and after such sales, investors kept companies like Harris's afloat with venture capital. The investors expected that these new internet-based companies would not turn a profit and would be rather expensive in terms of their quarterly losses (Cochran et al. 2006).

The promises of golden egg-producing geese were so compelling that many investors were willing to sign agreements that prevented them from selling their shares for a set period (Teeter and Sandberg 2017). Like Howard Hughes and his Spruce Goose, businesses built on soaring narratives sometimes fail to take flight. After several years of failing to secure funding, Harris's corporation filed for bankruptcy in 2001 (Salkin 2009). After the demise of his company, Harris moved nearly as far as a person could get from Silicon Valley and bought a small compound in Ethiopia in 2007. There, he seemed to make his life into one of his performance art projects, spending his days and nights musing mostly to himself, or to biographers who came to see what had become of one of the first internet millionaires. Who he had become was something of a parody of Martin Sheen's character in *Apocalypse Now*. It was only after losing his newfound millions that Josh felt the urge to disconnect utterly from the world he knew, to descend into a fractured version of his former self, sweating in a dingy room far from home (Smith 2012). Ethiopia was the perfect place for a failed internet tycoon to run away from the web he had helped to spin; less than half a percent of the population of Ethiopia used the internet when he arrived in 2007 (World Bank, 2020).

While Harris has since returned from his self-imposed exile, to help produce a documentary about himself, his story shows us several important dynamics in the contemporary internet. First, and one that must remain in the foreground is that the dot-com era was one in which corporate actors grew quickly in size and influence on the web. The second is that this era resulted, as in the case of Harris's company, in crashes that would mark the beginnings of feudalization in the contemporary web. As Matthew Crain notes, many investors bought into or sold many dot-com companies on the promise of a new idea with some future profit potential (Crain 2014) As if taking a note directly from Baudrillard, there was no longer any need for there to be a real company. Investors were more than happy to buy and sell companies based wholly on promises of future digital realities. This speculative investment in digital companies, aside from producing fascinating stories, also generated a bubble in the economy. Stories of personal interest captivate people, whether those people are readers of fiction, day traders, or managers of venture capital. In the dot-com era, the stories sold a massive number of stocks based on little actual ability to produce the promised goods and services.

Economists such as Crain, writing in the aftermath of the soon-coming crash, found that the trading on the mere stories of companies showed that investors of all kinds were trading on feeling rather than reason: without a history of providing goods and services upon which to base decisions, investors bought into dot-com companies based on the information they did have, which consisted mostly of the ideas and presentation of the owners and employees of the dot-com companies, as well as generally positive hopes for a newly commercialized internet's capability as a new frontier of consumer commerce. The dot-com crash in 1999 and early 2000 crushed those feelings, as well as many dot-com companies, resulting in dramatic changes to the online landscape, forming the flows of power and information on the contemporary internet.

BEGINNINGS OF FEUDALIZATION OF DIGITAL SPACES

The previous section's history of the early internet and its precursors, which centered on the role of the state in its creation of the ARPANET, ended in the early 1990s with the opening of the internet for commercial use. In that period of digital history, the state, through federal funding constraints, was the dominant actor in the actor-network, as the government limited access to the network to those engaged in the state's research projects. These early funding restraints, in light of the adapted-ANT analysis I proposed in the last chapter, were programs of action that shielded most people from internet access at all. With the opening of the internet to commercialization, however, corporations had the opportunity to rewrite those programs in ways that would begin to affect the agency of internet users as people came online beginning in the early 1990s.

Here, I take up more contemporary internet history to make the case that, though there were attempts to create spaces online aimed at building solidarity and mutual empowerment to the end of more democratic participation on the part of everyday people, corporate actors have, in effect, feudalized the internet, becoming the dominant set of actors in the contemporary internet. I focus on the concept of feudalization in digital power dynamics for two main reasons. The first is that in the vain hope for an internet that is hospitable to democracy, the ability for persons to communicate with one another through the internet makes it possible for people to deliberate with one another to empowering, democratic ends. It is possible, I argue, for participants in digital spaces to work together toward something like Habermas's ideal speech situation. Were the power dynamics in digital spaces such that people entered the actor-network on equal footing with other actors, then the internet might be just the set of technologies necessary to realize the Habermasian ideal of free speech. Keeping this optimistic potential in mind, I make the case here that, given a second reason for focusing on feudalization, digital power in the contemporary era more closely resembles that of a Habermasian take on feudal power. As Mark Warren explains in his analysis of Habermas's conception of power under both feudalism and liberal public spaces,

> If feudal politics meant legitimation by authority, the principles of liberal public space are different: Legitimation stems not from the authority of persons, but from the exercise of power have been put to the test of public discussion. (Warren 1989)

Taking Habermasian thinking into the context of the internet, if the "authority of persons" that marked feudalism becomes the authority of corporations, then the exercise of corporate power online may well foreclose democratic possibilities online if the platforms over which corporations exercise their authority do not privilege deliberative norms. Further, Warren notes that to reach the ideal speech situation in interactive contexts, participants must commit to a shared norm of persuasion through argumentation; to have a public in the Habermasian sense, we must enter a space as participants who seek to persuade and are open to persuasion by the speech acts of others, not by some other, strategic ideal, such as profit (ibid. 520).

The opening of the internet for commercial purposes, necessary to bring the web to people's homes, ushered in a new dynamic online. In the mid-1990s, with the help of graphics-based browsers such as Mosaic, people who were not experts in any coding language began using the internet more readily. As Mosaic, and later Netscape, America Online, and other browsers proved, people were more than willing to use the internet for all these things and more but lacked the technical skills to design and implement the necessary digital platforms to facilitate those activities on their own. This entrance of ordinary people onto the commercial internet as users additionally enabled a feudalization of the internet around corporate interests.

THE WORLD WIDE WEB

So far in this work, the network that would eventually become the internet has held several names such as ARPANET, the Web, the Internet, Web 1.0, 2.0, and so on. While all of these describe the internet, most of them have fallen out of common usage with "the internet" and "the worldwide web" remaining in common English language usage. With all these names in mind, it seems prudent to spend a moment on the role of naming in the construction of this network. For much of the recent decade, internet users, and vitally, companies, used Web 2.0 as the moniker for this actor network. Web 2.0 does not have a specific launch date in the same sense that, for instance, a sequel to a film containing the number two in the name will have a release date. There was no official announcement as there is no authoritative body to officially announce such things for the web. Instead, this new name for the web was a popular designation adopted in discourse retroactively to describe the post-dot-com crash internet. Web 2.0 was not a technical designation but instead was a marketing decision made by the companies that survived the bursting of the dot-com bubble of the late 1990s and early 2000s in an attempt to distance themselves not from the technologies of that era, which they were happy to keep profiting from and innovating on, but from the practices that nearly derailed the project of the internet as it was just beginning to reach more and more American homes. It is in the aftermath of the dot-com crash, I contend, that the landscape of digital power would shift, favoring a consolidated set of corporate actors who view and treat internet users as both consumers and producers of data. This powder dynamic privileges corporate actors' search for ever-increasing profits over the potentially empowering and democratizing communications capabilities presented on an internet that is accessible to large swathes of the population.

The brainchild of O'Reilly Media, a marketing company that specialized in dot-com companies, Web 2.0 became the moniker for a series of conferences beginning in 2004 (Musser 2007). These conferences demonstrate the state of power in this new era of the internet. The first of these conferences included speakers who continue to dominate the contemporary internet in economic terms, such as Mark Cuban and Jeff Bezos (Medialive International 2004). The former, speaking at the conference, suggested that the attendees should all make sure to oppose legislation that would, if passed, hold software developers criminally liable for crimes committed using their software (Olsen 2004). Their discussions, it seems, focused on developing the concepts of Web 2.0, covering themes such as marketing their new platforms, whether Web 2.0 was just another bubble and burgeoning social networks; the conferences focused on the internet as means by which to conduct commerce.

As the marketer behind the conferences, Tim O'Reilly notes in a germane blog post, Web 2.0 was a shift in how companies thought about the internet and the innovations that would realize those new conceptions. He admits that there were no release dates for the 1.0 or 2.0 versions of the Web but offers the IPOs of Netscape in August of 1995 (Shinhal 2005) or Google in August 2004 (Ritter 2014) as potential dates at which to begin the Web 2.0 era. The shift to 2.0, O'Reilly claims, was a "gravitational shift" from a model that considered the internet as a means by which to sell products and services to a model that considered the web as a platform in which companies could increase their footprint, bringing users into their platform as consumers and producers of data. In terms of this work, that shift recognized that the internet is a space in which actors come together to interact with one another through relations marked by power dynamics. O'Reilly's piece sketches the shift to

Web 2.0 as a change in the relationship between corporate actors and internet subscribers, in which the latter were no longer simply customers but vital producers of data, which is a key commodity on the contemporary internet.

Under Web 1.0, corporations saw everyday people as customers. Amazon sought to sell you books, Pets.com sought to sell you dog toys, and Stamps.com offered postage supplies. Under this Web 1.0 model, the economic relationship was the same as that of a person entering a brick-and-mortar establishment. Under Web 2.0, the product or service sold by the site is of secondary concern. Rather than being mere retailers, these corporate actors began to see, and market, their sites as platforms. On these platforms, engagement and interactivity between users were the core of the business model. At the 2010 version of the Web 2.0 Conference, for instance, Mark Zuckerberg spoke at length, making the case that the user was critical to the survival, growth, and profitability of Facebook as a platform. Users did not simply buy services sold through advertising links, but their internet behaviors also generate engagement and data (O'Reilly Media 2010). Both engagement with the products and services offered, as well as interaction with other users, generates data. A recent investor press release by Meta, for instance, indicates that the company's ad revenues, which make up over 90 percent of the revenues, have held steady at around $30 billion per quarter in the period from 2019 to 2021 (Meta Platforms 2022). Under Web 2.0, Amazon is as interested in selling data as it is in selling household goods, apparel, or books.

Additionally, O'Reilly adopts language that, read through the lens of this work, speaks to ANT as a methodology appropriate to the study of the web. Referring to Web 2.0 as a "sense" that was "formulated," the marketer nods to the fact that the internet is a social construction on behalf of its participant actors and that those actors attempt to implement programs of action into the network to shape the experiences of other actors. If this formulated sense takes hold, he notes that it will be key to the survival of Web 2.0 businesses to iink to one another and make use of the data generated by its users: the creation of a network of overlapping and complementary platforms will make 2.0 more resilient and adaptable in his thinking. This cooperative model, made possible in no small part by the participation of everyday people, speaks to Web 2.0 being, quite consciously, an effort at speaking into being an actor-network. As O'Reilly puts it in his description of Web 2.0, "in others, the winner will be the company that first reaches critical mass via user aggregation and turns that aggregated data into a system service" (O'Reilly 2005).

In the shifting nomenclature around the internet, we see the growing influence of corporations on this changing actor network. What began as a government project became, in the early 1990s, the domain of corporations who made clear and deliberate use of their power in this network to rename it in an attempt to not only advertise their supposedly new business practices to the public but, in my analysis, reify that power in rewriting economic and digital history in real-time.

DOT-COM BOOM AND BUST: CONTESTATIONS IN DIGITAL POWER DYNAMICS'

By 1994, barely two years after the web had launched for consumers, online stores were popping up, first to sell books, and then nearly anything under the sun, including a thriving black market that would catch the attention of several nation-states. For instance, the FDA and its British counterpart have been trying, largely unsuccessfully, to regulate the online sale and distribution of drugs, both prescription and nonprescription, since the mid-1990s (Walsh 2011). These attempts by the government to regulate online sales and distribution of drugs, in terms of ANT, are telling. An actor, the state, which had enjoyed large sway over online life in the era of the ARPANET, recognized that its hold on the digital tools it had helped create was loosening, and in response the state began using the legislative pen and the search warrant, to gain some part of that control back. In a similar vein, the 1996 Communications Decency Act was, in part, an attempt by the Clinton administration to bring the internet into the fold of state regulation via the Federal Communications Commission (FCC). If the courts would hold that the internet was functionally like radio and television broadcasting, the FCC could fine and censor websites that posted, shared, or hosted content that the FCC found to be indecent.

At the time, this brought up substantial concerns around the freedom of speech: since it was mostly persons, rather than broadcasting companies, who made and posted most of the content on the internet, would the act, as Charles Nesson and David Marglin speculate, suppress free speech on an internet that in the eyes of many, would foment an increasingly accessible and empowering norm of free communication between people (Nesson and Marglin 1996)? In 1997, the Supreme Court found that the act was overbroad, poorly defined, and, in the opinion of Justice John Paul Stevens, placed an "unactable heavy burden on protected speech" that "threaten[ed] to torch a large segment of the Internet community" (ACLU 2017). While the state would later attempt to regain some semblance of the control it had under the ARPANET, the 1990s would remain a period relatively free of state control online when compared to our more recent era.

For some, the ability to use the internet to challenge norms presented an opportunity to challenge practices to empower themselves to participate more fully in economic life. For others, these challenges allowed for the construction of new communities aimed directly at building networks of support, solidarity, and belonging, as well as spaces for civil disagreement and dissent, the very foundations of democracy. Through the commercially accessible internet, members of the LGBTQ community created forums, chatrooms, and listservs that were more welcoming than much of the offline world. In those more welcoming digital spaces, people could share their experiences, build interpersonal relationships, and craft communities that were not bound by immediate geography (Koch and Shuckman 1998). The internet, in this context, was a set of tools for building a community of shared experiences and identities, the material presence of the internet to link people together facilitated community building. Physical proximity, for members of the LGBTQ community, meant in the 1990s (and certainly still does today) visibility that can lead to immediate physical danger (Curlew 2019). The web, in its displacement of the immediacy of the physical body and lessening the need for immediate physical proximity for community building, made it easier for these actors to form networks for the explicit purposes of self and group empowerment (Miles 2018).

Communities of color have similar stories. A small but growing number of African Americans, whose interests in technology developed in the years before the commercialization of the internet, became part of the pre-internet technology sector. Many of these, for example, the human computers employed by NASA, made popular in the film *Hidden Figures*, came into the computer industry as specialized workers. These folks, the first generation of whom trained at technical schools, junior colleges, and Historically Black Colleges and Universities (HBCUs), were also among the early adopters of the internet. One group, who Charlton McIlwain terms "the vanguard," saw that first email chains, then more robust online forms such as Afronet could serve as important sites of empowerment and solidarity (McIlwain 2020). These forums, where people of color could come together to share experiences, chat, discover new uses for the newly opened internet, and perform aspects of identity that had otherwise been marginalized through, for instance, code-switching in white-collar workplaces, eventually became vital for their members. One facet of digital communities becomes vitally important here. In forums, especially small ones, moderators and members gatekeep; they can decide what is permissible within a given space and remove interlopers or abusers.

These early examples from the Black and LGBTQ communities show that the internet can certainly be a set of tools people use to empower themselves and others. But, I argue, the structures of power online that existed in the dot-com era have intensified, increasingly favoring large corporate actors' ability to dominate and pave over digital spaces to the detriment of marginalized groups. If you decide to go online today, most of the websites I have mentioned so far are either defunct or are something like digital graveyards, riddled mostly with ancient links that go nowhere. Since the dawn of the twenty-first century, many of these sites have become defunct, replaced by something else in the current era of the internet, often referred to as Web 2.0. To get to 2.0, however, some major structural changes occurred to the digital landscape that would influence how we interact with and through it. Through the lens of an adapted-ANT analysis, there is room for some optimism about the empowering potential of the early commercial internet. Materially, the widening availability of the World Wide Web made it possible for new communities to form, and those agents who made those communities could, and did, devote efforts and resources to maintaining and growing in spaces where solidarity, understanding, and acceptance are predominant norms. It is important to remember, though, that actor-networks such as these new online communities are contingent, and thus will neither necessarily survive at all nor exist forever as the kind of spaces that they had originally been.

If it is materially possible and people are willing to do the work, it is possible even on a commercial internet to carve out spaces within the actor-network that explicitly aim at helping people feel empowered to enter and engage with social, economic, and political life in ways that they had not before the advent of the internet. In the early years of the commercial internet, corporations were experimenting with methods to gain digital power and to claim digital territory more completely. These methods were in their infancy, and the failures of many dot-com corporations show that they were only sometimes effective: those companies that would go on to survive the dot-com crash learned from those experiences, and their methods for the establishment of digital fiefdoms have improved over time. Here, a consolidation of profit in the companies that remained and prospered in the Web 2.0 era, as well as their innovations and intensifications in programs of action that increase profit, would begin to reshape the web.

Some websites and platforms directly attempt the empowerment or assistance of their users, and they represent hope for the remaining promise of further development of the empowering potential of the internet. They also offer a glimpse into both the cultures of Web 2.0s users and the problematic behaviors that sometimes manifest in contemporary online communities. I will give two brief examples here. The first is a mutual aid forum, the subreddit r/random_acts_of_pizza. As the name would imply, the purpose of the forum is to provide people with the opportunity to ask for and receive a pizza. With its thousands of fulfilled requests, I argue it is a great example of the sort of human solidarity that demonstrates the sort of solidarity that can be facilitated through open, communicative discourse on the web. The gifting of a pizza to a stranger out of kindness may be a small thing, but it shows that people who choose to coexist, even anonymously, can help one another and form a community. If they can avoid devolving into a tribalistic debate over toppings, or at least work out how to peacefully discuss whether pineapple on pizza is tolerable by a person with a reasonable comprehensive doctrine, then there is some hope that we can learn to talk about more politically relevant things, too.

Another, the African American Literature Book Club (AALBC), operates under a similar general premise as Amazon did: to sell books online. It has, since 1997, done so with a focus on promoting and platforming Black and African American authors, giving space to otherwise marginalized authors. In addition to selling books, there is also a forum for the discussion of Black and African American literature, mutual aid, and general discussions that are germane to the lived experiences of people of color. These services and forums bolster, through their funding, other efforts aimed at web accessibility and providing publishing assistance to authors of color. AALBC, then, is a site that takes advantage of the Web 2.0 idea of the site as a source and invites the user to not only consume by purchasing books but also become an activist to forward empowering ends. As a retail site, it works within the existing actor network of Web 2.0 to recreate some of the adaptations taken by large corporate actors, albeit to the ends of Black empowerment rather than mere corporate profit. Making use of the idea of site-as-platform, AALBC invites internet users to become not only users but to join in a space of cultural progress where people can share, celebrate, and advocate for aspects of their identities outside of an environment where the user's data is commodified by a corporate actor.

What r/random_acts_of_pizza and AALBC have in common is that they have managed to exist within the Web 2.0 environment and do so to ends that serve the interests of people not as consumers or users, but as members of shared communities: and there are many others like them. In the case of the pizza gift-givers, we see that even in a highly corporatized framework such as Reddit, itself an immensely valuable seller of data, it is still possible to deliberately empower others, even if is through strangers sharing and creating solidarity through pizza. The subreddit, though moderator tracking, allows for people to keep "score" of pizzas given and received. Many users have scores in both columns, which indicates that people engage with this community over time. AALBC, making use of the ability of retail sites to become platforms, has sought to become a centralizing site for resources and spaces of conversation that empower people of color to engage through and with literature that advances their careers and cultural lives. In this sense, AALBC is a spiritual successor to the likes of Noirnet from the early days of the web, with the adaptations of Web 2.0. Examples such as these add credence to the use of an adapted-ANT analysis; where some variants of classical Marxism might see relationships between actors (usually framed in terms of class) as antagonistic struggles, an adapted ANT is open to describing, in this case, examples where the users behind Noirnet and other sites deliberately resist the programs for action that corporate actors attempt to implement in online spaces. Contingency is critical to ANT more broadly and for normative possibilities for digital democratic deliberation in the future.

The examples I have highlighted as proof of the possibility of spaces of empowerment and cooperation on the early web certainly appear promising in terms of fostering democratic experiences. They were spaces where people came together, on an open and equal footing, to discuss issues and interests of common concern to the ends of self and group success, as well as to ask for help from fellow human beings. In terms of improving democracy and civil society to achieve inclusive and democratic ends, one hopes for exactly those sorts of spaces, and they have existed before. Jürgen Habermas found similar spaces in the German coffeehouses, British pubs, and French salons of the eighteenth century. In the German theorist's telling of that period, the development of mass commercial publishing foreclosed the possibilities for the development of bourgeois reading publics. Instead of critical reading publics that prized deliberation, mass publics that prioritized production and consumption grew with the expansion of modern capitalist economic structures (Habermas 1991). This culture of mass consumption, Habermas reasons, is not the birthplace of liberal democracy, but the field in which a feudalized political life has taken firm root. In this feudalized digital landscape, we relate to corporations as serfs did to their feudal lords—dependent upon them for access to the bare resources of the actor-network, while also contributing to the power and longevity of the feudal system by paying the lord his share of the crop.

Habermas, studying the beginnings of mass politics in the modern era, found that capitalism, which was at the time industrializing, proved to be fruitful ground for the seeds of publics more concerned with production and consumption than with deliberation. What I observe in the history of the internet I have thus far presented is similar: the same economic system that produced mass publics in the eighteenth century, and highways in the twentieth, is likely to continue to produce technologies that lead to the concentration of power in the hands of the state and corporate actors. Undergirding all of this, and providing much of the labor, be it industrial, culinary, or digital, has been Mass Man. Mass Man, the concept Habermas's close adviser Theodor Adorno developed (Adorno 1991) to describe the serfs victimized by this feudalization, are not active participants in political life as Habermas found in bourgeois reading publics. They do not come into political spaces as equal participants expecting to hear and be heard by other equals before coming to collective decisions. Instead, Habermas finds that those living in mass democracies experience politics mediated through the discourse of production and consumption. Mass media, through which most of us get our information, is a proprietor of information as a commodity. The responsibility of these proprietors is to their financial stakeholders, not to the shared development of reasonable public discourse. Politics, in mass media outlets, is another product to be bought and sold; these outlets will just as gladly sell unverified supplements and pillows rather than, or often beside, the commodified opinions of pundits and other elites (Habermas 1991, 164). Politics, to mass man, is a product consumed with morning coffee in the privacy of one's own home before the day's production begins at work. This does not bode well for the promise of liberal democracy as Habermas sees it in *The Structural Transformation of the Public Sphere.*

Corporations, which cooperated with the state to build the networks that would become the internet, have a clear motivation. To corporations, profit is the motivation that drives their actions, and corporations, thus, build programs of actions into artifacts, such as computer networks, which are intended to facilitate the generation of profit. Those profits came not only from the state directly, during, and after the ARPANET period, when corporations were the contractors who built and maintained the network, but from the third set of relevant actors once the internet went commercial: people who use the internet, whether paid subscribers or otherwise. These internet users, of course, have varying motives for interacting through the web. Those I have followed so far in this chapter tried to use the internet to empower themselves and others by using the government and corporate-created internet to find and make communities in which that empowerment could develop. What the internet was then, and became later, depends on the balance of power and interactions between these actors. In accepting the commercialization of the internet, the state took something of a background role in the dot-com era: this, combined with the overall trend toward corporate deregulation in the decades leading up to the opening of the commercial internet, hastened corporate actors' taking power in the early web (Meltzer 1988). This stepping back of the state in the actor-network of the internet may have been, structurally, like Habermas's understanding of the waning of the feudal order in eighteenth-century Europe. With the state no longer dictating publicity as completely as it once had, there was room in that actor-network for new actors, bourgeois reading publics, to emerge and begin to operate to their ends. Similarly, the early online communities I have described existed at least in part because the state did not regulate the early web against it, and corporations had not fully taken over the agenda-setting and gatekeeping roles that they would occupy in more recent digital power dynamics.

In Habermas's analyses, read in terms of my adapted-ANT analysis, the private interests of the mass presses once made material via the publication of papers as a consumer commodity, were programs of action that affected the eventual content of publications and discourse. Making use of that space, they began to increasingly constitute the informational life of the public in terms of consumption. In other words, companies structured the network of eighteenth- and nineteenth-century European states such that it was more hospitable to enter it as a consumer or producer than as a citizen (Habermas 1991). The bandwidth of that network became increasingly devoted to production and consumption, space which was then less able to fill with discourse and action aimed at empowering everyday people to act outside of corporate spaces.

On the early web, corporations sought to buy up as much digital real estate as they could and offer the internet as a service, much like the press that had commercialized two centuries earlier. As the corporations acted following the motive of profit, they also began to alter the network: the results of these structural alterations shaped the next era of the web.

In the two centuries between the re-feudalization of European political life and the commercialization of the internet, critical scholarship documented the tendency of corporate, private, and capitalist actors to make use of and participate in the development of actor-networks with the cooperation of the state to create and expand their influence and power.

THE DOT-COM CRASH AS A WAXING OF CORPORATE POWER

As the turn of the new millennium neared, several events converged in such a way that they would both shake the foundations of the internet's industries and restructure the corporate actors that dominate the contemporary internet. Investing practices earlier in the 1990s began to shift, a recession began, and dot-com companies either adapted or closed. These adaptations would have substantial ramifications for the structural future of the web in terms of power.

Early in the dot-com era, venture capitalists were all too willing to throw money at new businesses based on interesting-sounding ideas for new digital services, with little to no solid proof that these companies would produce any such thing. This willingness, when contrasted to the Y2K scare, seems a jarring transition but one that demonstrates the power of narratives in online discourse to shape the behavior of actors within the actor network (Goodnight and Green 2010). Y2K was one such narrative. The most reasonable version of the "year 2000 problem," or Y2K, was a concern that various computing systems, ranging from airlines to banking, would not know what to do with data points once the clocks rolled over to 1/1/2000. The fear was that this would lead to widespread failures of computer networks in a global economy that was increasingly dependent on networked computers. In less reasonable versions, some Americans feared that the ball dropping on New Year's Eve was a signal to world governments to declare martial law, shut down telecommunications, and begin a new world order.

Some shareholders and investors became increasingly reluctant to invest in technologies that they understood poorly. That concern would become magnified when, in March of 2000, Japan's economy—a major source of computers and software—entered a recession. With these concerns and the recent expiration of the employee and investor agreements to hold dot-com company stock for set periods, a global selloff of dot-com stocks began, with scores of companies left valueless, shells of their former selves if not closed altogether. There were some large corporate casualties in the crash. For example, Pets.com, which had the marketing budget to purchase a Superbowl advertisement in 2000, went defunct shortly thereafter (Thornton and Marche 2003). The same is true of E-stamp, a stamp retailer attempting to make its way through the digital age that had dumped most of its IPO funding into advertising in a bid to save itself (Crain 2014). This is just one small measure: businesses in nearly every sector one could imagine shared similar fates. The dot-com crash, I argue, cleared many of the actors from the early internet from the network entirely, allowing for others, in the aftermath, to grow in terms of capability to influence digital power dynamics going forward.

Many of the survivors of the dot-com crash pursued a strategy of enlarging themselves to survive in this post-crash environment. The clearest example of this is America Online. By 1999, AOL was a company in search of one thing it did well. It was part service provider, part media company, and part email host. Keeping that corporate ship afloat became difficult and media conglomerate Time Warner purchased AOL, infusing the latter with capital for several more years. Despite concerns that this merger ran afoul of the spirit of antitrust laws intended to keep individual corporations from gaining too much power in economic life, the sale was allowed and AOL, which was spun off in 2009, continues to operate today as a media aggregating company with email service (Aufderheide 2002). The AOL sale was the first and clearest example of what would become a trend in both digital economics and power structures, the rise of giant companies that would hold effective monopolies in their market segments. Although there was scholarly, regulatory, and journalistic concern that the AOL–Time Warner merger would lead to a monopoly in the communications industry more broadly, the Federal Trade Commission allowed the two media giants to merge, providing that the combined entity assist other, non-affiliated broadband services in becoming available to potential customers before selling AOL's broadband subscription to customers in a given location. Upon their merging, however, the combined conglomerate became a one-stop shop for media consumption. In 2003 one could watch television, surf the web, send emails, and consume advertising from any screen in one's home or office having transacted with a single financial entity that now held a large swath of digital territory.

There are other examples of corporate giants that come immediately to mind: Amazon, which nearly failed for lack of funding in the early 2000s (Davis 2018), and which survived the crash, now handles nearly 90 percent of all web search traffic (Fishkin 2018) and is among the most profitable companies on the planet. Additionally, there are around one and a half billion users of Gmail (Elias and Petrova 2018), Google's email service, meaning that there are more Gmail users than there are people in China. These few companies mark the beginning of a structural pattern that emerged in the post dot-com crash era: giant corporations began to have near-monopolies over certain features or facets of online life. Here, I would like to return to the highway discussion from earlier in this chapter.

As I took up earlier in the chapter, the twentith-century highways were actor-networks that, at least in part, empowered states to expand their power and capabilities to influence the lives of people within those states. Similarly, as companies began to develop near-monopolies in segments of the post-crash internet, I argue in the remainder of this chapter that they expanded their power by regulating the behavior of the people who make use of the web. The communities I mentioned at the beginning of the chapter have been, much like Overtown, paved over. Instead of the federal government working in cooperation with a county commission, corporations working in the wake of ARPANET paved over other possibilities for digital life. The territory may be digital, but the effects are similar, communities now exist under the digital overpasses, the derelict sites from decades past in the shadows of a glittering, homogenous platform above. Whether digital or physical, this also forecloses other possibilities. It is, for instance, difficult to host a citizens' meeting on I-95 unless one is willing to deal with state and local law enforcement. It is just as difficult to build communities of empowerment when so much of the web consists of the traffic of Amazon, Google, and the like. On the contemporary internet, economic and geographic barriers to access, as well as to creating and participating on websites, place people working outside of corporate spaces at a relative disadvantage in their ability to make room for themselves, and each other, on the feudalized internet.

Put differently, an internet designed by the state and corporations is almost certain to work better for the state and corporations than it does for other sets of actors. By the end of the dot-com era, the internet as we know it now had begun to appear. With the commercialization of the internet a decade before, the state had taken a step back, which allowed corporations to take a large role in influencing online life. In the time after commercialization, corporations are adapting, or attempting to adapt to, the digital environment. At a time when corporations had not yet found their evolutionary strategies to ensure their digital hegemony, there was some room for empowering communities to develop. The dot-com crash foreclosed much of this development, after which the development of massive digital companies carved out increasingly large online fiefdoms. This feudalized internet, now called Web 2.0, began in earnest; the lines along which that feudalization took place is of little surprise if we hold that corporations imparted capitalist effects as programs of action into the web from its commercial inception.

THE INTERNET AS A FEUDALIZING ACTOR-NETWORK

As I demonstrated in this chapter, the state dominated early networked computing. In the same historical period in which the state developed highways to facilitate defense, it also, using similar means and to similar ends, developed the precursors to the internet. But the state was never the only relevant actor online; the same corporations that had developed the infrastructural bases and software of early, state-dominated networks sought to use those networks to the ends of profit beyond government contracts. Thus, those corporations, and others, took full advantage of the early commercialization of the internet in the early 1990s to dominate as much digital space as they could. This era, now commonly called the dot-com boom, saw a relative waxing of state power online as corporate power waxed with the growth of digital businesses. These businesses required customers, and it was as customers that ordinary people first came online. Even in the dot-com era, corporate interests, namely profit, condition much of peoples' online lives, which, in my view, present substantial barriers to the internet being a public in Habermasian terms.

The dot-com era, which ended not only with an economic bubble burst but also a deliberate attempt by corporate actors to rebrand in the aftermath of that burst, gave way to the contemporary internet in which the balance of power further developed. In the dot-com era, people entered the internet mostly as customers in the traditional sense, to buy or consume goods or services. On the contemporary internet, however, people are not only consumers but producers: the data that users produce is the most valuable commodity of the twenty-first century. Thus, corporations now have redoubled their interest in keeping people online as prosumers. This relationship, in which corporations act as feudal lords over domains in which we spend increasing proportions of our waking time, affects the agency of users. Since the commodities produced and consumed in this feudalized actor-network are time, data, and attention, there is no specific need for the information shared through digital platforms to be truthful. So long as data is produced and consumed nearly all the time, users will be affected and continuously trained as prosumers. It is that effect that I explore in the remainder of this work.

Chapter Two

Compressed Political Temporalities in a Feudalized Internet as a Threat to Peaceful Democratic Participation

The preceding chapters of this book drew a critical history of the internet through the lens of an adapted-ANT analysis influenced by the descriptive elements of new materialist thought. In so doing, I made the case that the contemporary internet, in its feudalized state, forecloses some deliberative possibilities through the deployment of capitalist programs of action by the corporations that own online platforms. This foreclosure represents a feudalization of digital politics that envisions the corporation as the actor with hegemony over much of digital life and inscribes people as producers and consumers to the detriment of more potentially democratic digital life. In this chapter, I turn more closely to people's experiences with technologies that hasten the flow of information and make the case that, in its current configuration, the speed at which the internet provides information to people as producers and consumers place people at the edge of their cognitive limits as human actors, making it more difficult to think and act critically to democratic ends. To understand how otherwise reasonable people would take to political violence spurred on by publicly acknowledged lies, I explore Paul Virilio's work as a means to better map the effects of digital, prosumer politics on people's minds in terms of their perception of time and political importance.

In describing contemporary society's obsession with technological progress, French architectural and social theorist Paul Virilio comments,

> That progress is nothing more than the progress of a deliriously bustling eagerness, not to say a collective rage, triggered by a sudden panic that's turned into a PANDEMIC. A pandemic that has everything to do with the reality effect of the

> acceleration of information and its sudden demands. For, the INSTANTANIETY of the disarray of each one of us will soon contaminate the way of life of all. (Virilio 2012, 61)

Though Virilio was making use of the word "pandemic" as a metaphor describing the realization of terror and deterritorialization of social and political life in the contemporary age, it has come to pass that we have faced a literal pandemic on a global scale. This pandemic has occurred on a similar scale; then it spread to the internet and easily affected human lives. The effect he describes, disarray, has taken hold of our political lives and its contamination has become apparent as the institutions of democracy have fallen under attack in recent years. This nearly prophetic statement by Virilio places its fingertip on the pulse of some of the most pressing and troubling aspects of digital political life, and they demand careful examination if we are to better understand how power functions in an increasingly digitized world in which a feudalized internet shapes how corporate actors, people, and the state interact.

This chapter turns directly to developing an understanding of the processes by which user experiences on the feudalized internet erode people's ability to think critically and (therefore) participate in democratic politics. To do so, I focus here on the ways in which people experience time through technology. Making use of the work of Paul Virilio as a jumping-off point, I make the case that many aspects of the corporate-centric, feudalized internet cause users to experience task saturation. This task saturation, which I further develop in the latter half of this chapter, entrenches people in the modes of thinking that they previously inhabited and limits the users' ability to evaluate new information critically. Through this entrenchment, and in a state of nearly continual digital production and consumption in a feudalized web, I find that some participants in democracy face threats to their ability to carefully think through their actions, sometimes to disastrous results. To these ends, making use of the political implications of the psychological work of Daniel Kahneman adds nuance to the adapted-ANT analysis that I present here.

For some internet users, this has meant the acceptance of falsehoods, which has begun to not only foreclose the possibility of democratic deliberation but spurs the transformation of some users into anti-democracy combatants. The existence of a feudalized internet alone does not adequately explain why some users have chosen, based on their online experiences, to try to overthrow their government through violent means. To further account for the increasing problem of the spread and acceptance of political falsehoods, I take up the effects of time on political reasoning. Through analysis and expansion of the work of Paul Virilio, I make the case for considering time as a vital dimension along which we can understand political behavior and that the temporal compression we experience through the contemporary internet disempowers users from some basic capabilities of democratic participation. Corporate platforms, the center of many users' digital lives, compress the experienced temporality of people who produce and consume data and content on those platforms. In an actor-network in which corporations make their profits off of continuous engagement, people receive and create political information in a compressed state, which is a far cry from the careful, thoughtful deliberation that Habermas hoped for.

This compression has made it such that users increasingly attach themselves to whatever sounds true based on what they already believe, while having less time to critically evaluate new information, which reduces some people's capacity for deliberation while increasing their willingness to leave deliberation behind for a reflexive politics that has turned, in some cases, to violence. In short, the wells of power in the actor-network of the contemporary internet serve as sites for radicalization and extreme politics. This radicalization has resulted in a destabilizing of the norms and institutions that are at the foundation of democracy by making it nearly impossible for users to take the time to think through their political choices and actions. As a main example of this, I walk through the spread of fake news to the point that it resulted in an attempted overthrow of the U.S. government in January 2021. To develop this critique of digital politics, I begin by analyzing the work of Paul Virilio at his most directly political.

SPEED AS A POLITICAL FORCE

Paul Virilio, a French architect, cultural critic, and self-described urbanist was one of a group of thinkers who were searching for explanations of power that would explain the speed and scale of the destruction wrought by states in World War II. Here, Virilio's intellectual circle overlapped with that of Michel Foucault, and he counted Gilles Deleuze, and to a lesser extent Felix Guattari, among his friends and collaborators both intellectually and politically. Virilio was present at the time and place of some of the most rigorous development of post-structuralism, though he often found himself at intellectual odds with Foucault and Baudrillard in particular (Virilio 1999). He found himself, then, as one of the cultural theorists who, like his contemporaries, sought to understand how power works and would work, in the postmodern world.

There are several concepts that Virilio developed over his career that are germane to developing an understanding of the effects that the temporal dimension of the internet is having on the political lives of people who use it. His engagement with disembodiment and deterritorialization through the experience of speed informs an understanding of people who use the internet as people constituted not only in space but, vitally, in spacetime. This constitution in spacetime adds another dimension along which we can understand the empowerment of certain actors in the network and the disempowerment of others and adds complexity to the sketch of the digital actor-network in terms of actors' asymmetrical positions of power within it.

Bringing Virilio into conversation with others thinking about how technologies mediate people's interactions with one another is more than worth the intellectual wrestling that Virilio sometimes demands, as the temporal dimension of our digital experiences is of vital importance to understanding the conditions under which people become not only prosumers but also political actors who have been left at a disadvantage when attempting to discern truth from lies that have led some of them to commit acts of violence against the state and their fellow countrymen.

Among the concepts that most often appear in Virilio's work, and are most relevant to digital politics, is his understanding of speed. Having witnessed, as a child, the German Blitzkrieg, a lightning war that relied on surprise, rapid-fire attacks by overwhelming forces, Virilio devoted much of his work to the centrality of speed as an element of strategy. Speed, for Virilio, has always been a part of the military strategies taken up by combatants. For example, a castle's walls were not merely spatial elements, but temporal ones as well, meant to alter speed. Those walls slow down an attacker's ability to harm those inside the walls, making a siege a matter of speed: the victor will be the one best able to cope with the effects that the walls have on the rate at which supplies dwindle. Speed, thusly, is a concept inexorably tied to territory.

In the medieval world, the walls fixed the site of the battle and slowed the tempo of battles, making them into sieges. This became inverted in World War II, Virilio argues, when the Germans used their new mechanized units to simply drive past France's impressive but immobile Maginot line; the speed of the warfare rendered the territory of the walls useless. The offensive strategy of the Germans was to overcome the relative lack of manpower with sheer speed and in so doing rendered the well-prepared but fixed defenses of the French obsolete (Virilio 2006). Here, writing originally in 1977, Virilio stakes one of his main claims, that with World War II came a shift in the strategy of the state from a strategy that focused on territory to one that focused on time to overcome the physical limitations of that territory. Speed has been and continues to be a strategic tool by which states could attempt to achieve their objectives on physical territory. One can then read strategic history as one of temporality. What took months or years in the ancient and medieval world of sieges and days in the Clausewitzian horror of World War II would take minutes in the Cold War, and perhaps seconds in the not-so-distant future. When time became useful as a tool to compress space, the territory became less of a focus in the formulation and accomplishment of strategic goals. The result, he describes, is that

> [p]rogressively doing away with our awareness of distances (cognitive distances), speed, in its violent approach, distances us from sensible realities; the more rapidly we advance toward the terminus of our movement, the more we regress until speed becomes, in a certain way, a premature infirmity, a literal myopia. Where the "lookout machine" [machine de guet] (spyglass, observation tower) brought the horizon close by the domination of altitude or the optical properties of lenses, the "machine of war" [machine de guerre] (vehicles, various vectors), in propelling the passenger towards the horizon, separates him to the point of being in an adjacent world, so much that we could consider the play of the vehicular proximity to be a detaching, a recoiling, a literal retreat. (Virilio 2007, 113–14)

Much of Virilio's work on speed, from *Speed and Politics* to *Pure War*, focuses on the state as its central actor; he does, at times, turn to other sets of actors. Here I am concerned with the actor who eventually inherits this myopia he describes in the passage. To this end, his "passenger," I surmise, is the group of people subjected to the accelerated actions of the state, and now corporate actors. Speed is only ever contextual: it is relative to a set of objects and observers. For instance, many learned of the speed of modern warfare once it came crashing through their village in May of 1940. We, those who are living through the feudalized age of the internet, are the passengers who experience myopia induced by speed.

The case for speed-induced loss of people's vision is a well-established empirical phenomenon understood in two interrelated phenomena: reaction times and task saturation. First, many readers may remember the following from their driving education courses: the faster one goes, the less time one has to react to an event a known or unknown fixed distance away. For example, at 100km/h, a driver has two seconds to react before reaching a point 200 meters down the road. At 200 km/h, it takes half the time to travel the same distance, necessitating longer follow distances between vehicles as speed increases, assuming all the drivers value their own continued lives. Once prompted, drivers can hit the brakes in their car in around three fourths of one second (Johannson and Kare 1971). That means that even before a driver takes corrective action, someone driving at 100km/h would have traversed about 75 meters, double that at double the speed. In this simple example, the relationship between the self, time, and territory is abundantly clear: the faster one goes, the less time, and less distance, one has to react meaningfully to a change in one's environment. That same three quarters of a second that our brains and bodies need to react gives us less time to think and react the faster we are traveling. But speed is not the only factor, as Virilio makes clear; we are often operating machines when we are in literal, spatial motion. When that is the case, another empirical phenomenon is at play: task saturation.

Conceptually Lt. Col James R Groff defines task saturation as "the state of having too much to do with too little time to do it all" (Groff 2006). The concept is one, keeping in conversation with Virilio's above-quoted passage, tied to the increasingly complex machines of war that states develop and field. It is, in Virilio's terminology, one of the integral accidents of the age of machines: train derailments came with the advent of the train. Virilio aptly points out that with new technological advances also, inexorably, come unintended consequences that often have ill effects on human beings (Virilio 2012, 45). Speed, in part, facilitates these accidents and determines their severity. The difference between a major automobile accident and a small cosmetic scratch is a matter of speed. With greater speed comes more severe consequences from the accidents that we have produced with the invention of new technologies. Considering these accidents in terms that center on the human actor, task saturation is a type of accident that disempowers otherwise competent human beings from being able to accomplish the ends that they had set out to do. Virilio, taking a similar track, finds the following:

> Temporal compression, as it is technically called, is an event that concretely modifies everyone's daily life at the same time. In this face of this acceleration of daily life, fear has become an environment, even in a time of peace. We are living in the accident of the globe, the accident of instantaneousness, simultaneity, and interactivity that have now gained the upper hand over ordinary activities. (Virilio 2012, 45)

People often experience task saturation because of temporal compression, and one possible consequence of these phenomena is fear. When we experience our lives at an accelerated pace, Virilio argues, we become fearful. In that state of fear, we lose the ability to think clearly and make mistakes. Those mistakes, or accidents, override our otherwise normal activities with their consequences, sometimes making the continuation of normal life impossible. In the more directly mechanical realms of driving or flying, crashes are the result. What then, are the consequences of task saturation in a political environment marked by a prosumer-based model of information production and consumption to the ends of corporate profit?

In terms of my adapted-ANT analysis, the disparities in power that I outlined in the previous chapters can be explained in terms of the agents' experiences with time. Whereas corporations can think in terms of their quarterly and annual profits from the data collected by users, the users experience the internet both constantly and instantly: the former state allows for thought, the latter might well lead to "the accident of instantaneousness, simultaneity, and interactivity" that Virilio describes.

CONSEQUENCES OF TASK SATURATION

Remaining for a moment with the operation of machines, examples of task saturation are plentiful. First, from the field of military aviation, the Air Force found that medical transportation teams that seek to evacuate patients via helicopter and provide critical treatment before arrival at a medical facility are task-saturated by their missions, involving flying an aircraft and caring for a critically injured patient or patients, nearly half of the time, resulting in adverse outcomes for (thankfully, simulated) patients in nearly half of missions. These crews, well-trained aviators and medical practitioners, simply by the fact that the situations with which they are presented move too quickly, do their jobs worse than they should nearly half of the time. For their patients, this failure rate is unacceptable, and corrective training typically means learning to rigidly follow premade checklists to maintain standards of care and safety (Davis et al. 2013). The psychological conclusion is clear: there seems to be a point at which human beings cannot engage critically well enough to perform in fast-moving situations, and in those situations, they turn to already-known ways of behavior to cope with the speed of a given situation.

Another example of a situation that is high-stakes, fast-moving, and accident-prone is the landing of an aircraft. In most cases, there are rigid checklists in place to ensure that the flight crew has the best chance of safely putting the machine, along with their passengers, on the ground. There are cases, however, such as a 2015 incident that killed the passengers and crew of a small flight, where two competent aircraft operators, flying a perfectly flyable aircraft in good weather conditions, landing a plane they were both trained on, managed to crash. The reason for this, the FAA finds, is simply that they chose to deviate from the checklist and reason through the landing, something either the captain or the first officer should have been able to do. They were, in simple terms, task saturated in a situation that moved too fast for their limited, human brains to process in logical ways (National Transportation Safety Board 2016). People, even highly trained ones who intend to complete difficult, complicated objectives experience task saturation when they become overwhelmed with too much information, too quickly for their brains to process with adequate evaluations of new and changing circumstances. When a situation confronts a person with too much information and a lack of time to critically think and process it, those persons, as these examples show, tend to fall back into patterns of behavior that they already know. The lesson, that even well-trained specialists such as pilots rely on ready-made heuristic devices such as checklists to do their high-stakes jobs in speed-compressed environments to combat task saturation that can result in negative outcomes, has begun to find its way into medicine, as well (Muller and Patel 2012).

Task saturation and speed-induced accidents are not the realms solely of specialists doing surgeries or flying aircraft: daily commutes provide more quotidian examples. About three thousand Americans die every year combining otherwise easy tasks: texting and driving (Centers for Disease Control and Prevention 2021). The empirical facts stand up well to Virilio's general hypothesis, that speed induces a myopia of sorts. When an otherwise normal situation (a trained pilot flying an aircraft, a surgical team operating, or a motorist driving a car) speeds up, we lose sight of the fact that each of these tasks is well within our means to accomplish and, without the aid of some premade tool, fall prey to what is essentially a cognitive lockup. Of course, no reasonable pilot would crash their plane, nor would a surgeon leave a tool behind, nor would we rear-end someone at a stoplight; these accidents result from a person whose brain cannot adequately process and think through several tasks at once.

DIGITAL TASK SATURATION

For people who are participating in the consumption of news, entertainment, and other information online, the immediate stakes of task saturation are less mortality-inducing, assuming the person is not doing so while driving. Instead, I suggest here that the same phenomena of task saturation and reliance on heuristics occur with users of online platforms; these platforms expose people to more information than they can critically evaluate at any one given time, and, thus, result in other political heuristics that offload the work of interpreting the meaning of information to some other actor. In so doing, users become disempowered from the democratically vital processes of critical evaluation of arguments and new information. In terms of the broader work, task saturation, because of physical speed, creates a situation of disempowerment for the person who has become task saturated.

I turn now to the more direct political consequences of these phenomena in the age of digital democracy. Here, Virilio offers some commentary that is a useful jumping-off point for further analysis:

> In the speed of the movement, the voyeur-voyager finds himself in a situation that is contrary to that of the film viewer in the cinema, it is he who is projected, playing the role of both actor and spectator of the drama of the projection in the moment of the trajectory, his own end. (Virilio 2012, 106)

Again, some clarification is necessary to bring the somewhat elusive Virilio to bear on the question of the political importance of speed. This "voyeur-voyager" is, in my interpretation, the driver of a given vehicle. One of the more important characteristics of a vehicle, in Virilio's analysis, is the windshield or screen, about which he makes a cinematic comparison: on the windshield we see the world coming at us at high rates of speed, but we see it presented as a film in which we can only indirectly participate through the controls of the machines. What we see on the windshield, in the age of machines, is mediated by mechanical objects that put us at a distance from that which we are seeing unfold through the windscreen (Virilio 2012, 41–42). When operating a vehicle and looking at the world through the mediation of the windshield or screen, "the immediate proximity matters little, the only important thing is that which is held at a distance; in the continuum of the trip, what is ahead governs the progress, the speed of propulsion produces its own horizon; the greater the speed, the more distant the horizon" (Virilio 2007, 11). Here, Virilio comes to the same conclusion as that found in the examples I have given, albeit from a different perspective. For the French urbanist, speed has the effect of making the immediate surroundings less and less relevant. He found this historically in his work on strategy; the speed of the German offensive rendered the more physically immediate geography of the French defensive positions less relevant and useful to their designed ends. The operation of mechanical machines can, too, lead to disastrous consequences as in the cases of car crashes and the like caused, in part, by speed-induced task saturation. With the addition of speed we have, at the cognitive level, challenged and not frequently overcome the limitations of our human forms.

Virilio does not limit his concern around speed to the personal or the strategic realms, though both of those realms remain relevant throughout his work. Speed, for Virilio, is a concept that requires politicization:

> We must politicize speed, whether it be metabolic speed (the speed of the living being, reflexes) or technological speed. We must politicize both because we are both: we are moved, and we move. To drive is also to be driven by its properties. There is thus feedback between the two kinds of speed: technological (the car) and metabolic (man). There is work to be done which is tied with the vehicle, with the politicization of conduct in the Latin meaning of condurcere, "to drive," as well as in the sense of social conduct, of the conducting of war, of the economy. Speed is not considered important. Wealth is talked about, not speed! But speed is just as important as wealth in founding politics. Wealth is the hidden side of speed and speed the hidden side of wealth. The Two form an absolute couple. People say: "You are too rich," but no one ever says: "You are too fast." But they're related. There's a violence in wealth has been understood not so with speed. (Virilio 2008, 43–44)

In this passage, Virilio makes the assertion that lies at the heart of the critical interrogation of his work. When the metabolic speeds, represented by human reactions and behaviors, meet with technological speeds, the link between the two is clear and often consequential for human beings. It is along similar lines that, in previous chapters, I make the case for considering the internet as an actor-network in which the state, corporate actors, and people interact through technological means with one another in ways that are consequential to the balance of power in the contemporary world. Virilio begins, in this passage, to not only call for the work of making apparent the role of speed in contemporary political life but adds it as an addendum to the commonly held social scientific and theoretical subject matter of wealth.

The political implications of this have been speculated not only by Virilio but also in James Rosenau's *Distant Proximities*, in which he predicts that not-insignificant portions of human populations will see the changes of the digital era with fear and will fall into tribalism and nationalism to cope with the world that is rapidly changing around them (Rosenau 2003). People and groups have held ideas such as nationalism and tribalism as political heuristics since long before the dawn of the digital age. Following Rosenau's argument, but modifying it with the insights of Virilio, I argue that when faced with too much information, especially information that might threaten simplistic explanations and worldviews such as those often found in nationalism and tribalism, some people resort to their old heuristics, and retreat into their smaller communities whenever and however they can.

That a speed-assisted disconnection from immediate geography aided by digital technologies would result, as it has in the events leading up to the attempted overthrow of the United States government in January of 2021, seems on its face a paradox. How could making the state and its attachment to geography *less* important through speed result in *more* nationalism on the part of not only the state itself but people turned anti-democracy combatants? Again, consider the disparities in power and how those might affect the agency of actors within a given actor-network. If, for example, ordinary people found themselves exposed to information with every waking moment and left with little time to think through the validity of that information, might those people, over time, become less attached to the physical geographies that make up their lives and become all the more attached to the narratives that at least render their anxieties and fears some sense of coherence?

For Virilio, the relevant group of ideas that concern themselves with speed is part of a study he calls "dromology," or the study of movement (Virilio 2012). Before proceeding with more specific dromotological insights on the internet's effects on the ability of people to make sense of information, a few general principles can light the way forward for clearer analysis. Central among these is a reminder that we must keep in mind: speed, especially that generated by machines of any kind, is fundamentally and uniquely anthropocentric (Virilio 2006, 121). In conversation with the broader purpose of the work, this means that the speed generated by the internet can be attributed to human actors who themselves have intents and purposes; this makes dromology a potential line of thinking along which actor-network theory can be enriched. In much the same way that a materialist-influenced, adapted ANT expects asymmetry in terms of economic power, the asymmetries along which human beings, algorithms, and computers interpret, experience, and generate data will express themselves through different effects on each of those actors.

INTEGRAL ACCIDENTS

A second critical dromotological principle is that all technologies, whether the printing press, the car, or the internet, contained in their creation what Virilio refers to often as "integral accidents": with the invention of any technology that increases speed and involves human beings, the result for one kind of a crash or another is not only possible but made inevitable by the very creation of those technologies (ibid. 12). The only way to prevent all car crashes, keeping with the logic of the integral accident, would be to uninvent the automobile. Furthermore, to understand any given technology is not only to understand its original purpose set by those actors who wrought it but also to understand those integral accidents as well. Keeping these two dromotological principles, the anthropocentricity of speed and the integral accident, we can begin to interrogate Virilio's thinking on contemporary technologies in keeping with those principles to cut down the subject matter in his far-reaching works.

In his study of speed in the contemporary era, Virilio identifies that these "new" technologies are not, perhaps, quite as new as those hoping for the promise of technology may have us believe. For example, drones have become one of the technologies most critiqued by those concerned with contemporary warfare. Here, Virilio marries that concern to his strategic analysis of World War II, indicating that the Germans' V1 rockets were the first robots used in large numbers in combat (ibid. 31). That technology, which used the speed of flight to make the English Channel's geography, long key to British defensive strategy, less relevant when placed in a longer history when Virilio states:

> I am convinced that just as speed led to the Germans' incredible domination over continental Europe in 1940, fear and its administration are now supported by the incredible speed of real-time technology, especially the new ICT or new

> information and communications technologies. This technological progress has been accompanied by real propaganda, notably in the way that the media covers the new creations presented by Steve Jobs, Apple's all-powerful CEO. This combination of techno-scientific domination and propaganda reproduces all of the characteristics of occupation, both physically and mentally (Virilio 2012, 16).

Some nuance is necessary to connect the dots that Virilio paints so well for us here. The key to the analysis I provide here is that, for Virilio, the strategic use of speed is not the sole domain of the state. In the digital age, the state is in a nearly perpetual siege: when confronted with the end of the Cold War, it entered into new, global wars and has developed the surveillance state to pursue its new strategic ends of control and total panopticons (ibid. 21–22). This shift, I argue, is of vital importance to an understanding of power in the contemporary internet age. Second, of specific importance to this chapter, is the idea that these corporate actors make deliberate use of propaganda to occupy "both physically and mentally."

This disruption of an individual's general freedom of spatial and temporal movement is at the heart of his dromotological concern with the windshield: when we are moving at high speeds, we are never truly in any given spot for more than an instant. While this is true at any speed of movement, whether a car or an aircraft, it is vital here to remember that speed is a quantity: the faster that we move, the more pronounced its effects will be.

The internet, Virilio reminds us, is not likely to be different and will also come with its own set of integral accidents (Virilio 1999, 91) In the digital age, we do not experience politics at the speeds of cars or even jets, but at the speed of code being moved from one machine to another, which occurs very nearly at the speed of light (ibid. 82) If that is the speed of politics in the digital age, and speed is to be taken as a quantity, then the political effects of the web and its integral accidents can be expected, in a dromotological analysis, to be of the highest consequence possible. In inventing the internet, "we have reached the limits of instantaneity; the limits of human thought and time" (ibid. 33). Thinking in terms of speed, I also consider the rate at which integral accidents may occur, and their severity. An accident involving horse-drawn carriages, for instance, is likely much less harmful than one involving cars; adding jet aircraft to the mix certainly increases the chance of a crash that will affect many more people, more severely, and more quickly. More speed, weight, and size make crashes worse for those actors involved in them. If the speed at which information reaches human beings is the speed of light, then the integral accidents of the internet, including disempowering task saturation, too, occur at this high rate of speed.

With that severity of consequence, an analysis of political power in the digital age must include speed; it is the very speed of the internet that has generated the digital equivalent of car crashes. Thus, anthropogenic integral accidents are some of the defining features of politics in the age of the internet. In beginning to sketch those political features, Virilio finds that we have entered a new kind of politics, as he explains:

> Transpolitics is the beginning of the disappearance of politics in the dwindling of the last commodity: duration. Democracy, consultation, the basis of politics, requires time. Duration is the proper of man; he is inscribed within it. (Virilio 2008, 42)

Through his notion of transpolitics, his more general observation is that increased speed has given people substantially less time to process political information. I need not hazard a guess at the stakes of this in Virilio's work, as he offers the following on the matter:

> With the phenomena of instantaneous interaction that are now our lot, there has been a veritable reversal, destabilizing the relationship of human interactions and the time reserved for reflection in favor of the conditioned responses produced by emotion. Thus, the theoretical possibility of generalized panic. (Virilio 2012, 31)

Here, Virilio lays out more clearly the effects of the internet, of instant politics, on people who use contemporary technologies. These instantaneous interactions function much in the same way as the task saturation examples that I gave earlier in the piece. When we experience the temporal compression of digital politics, we lack the time to think and we are, in effect, task-saturated in the same way as pilots or drivers. Our conscious minds, which we used to reflect in political discourse, are not up to the task of doing so at the speed of light, and we thus rely on emotional responses, often panicked ones, as a poor substitute for the careful thought, reflection, and articulation necessary for democratic deliberation. It may not matter, then, that a platform allows for the instantaneous sharing of ideas between people if the fact that it is instantaneous means that people cannot contemplate what they are consuming and react emotionally to it.

In much the same way that a driver may well crash their car when they are texting behind the wheel due to task saturation, I argue, someone may well believe a news story because they do not have time to react critically to the new information that their phone is presenting to them or that they spend so much time on the web consuming such "news" that there is less and less time to think about the presented information critically. Wendy Brown, examining the American political landscape in the wake of the 2016 elections, reflects similarly to Virilio: the disorientation that some people feel in the neoliberal, increasingly feudal present leads some people, fueled by the rage that they find so ready for prosumption on social media, to lash out in nihilistic violence (Brown 2019). The perceived urgency of prosumed digital politics generates accidents through task saturation in much the same way that occurs in the physical examples above; left with not enough time to think, people behave in ways that they otherwise would not, sometimes to disastrous ends. In an actor-network where people interact as prosumers, there is little to discourage the conditions of task saturation: if anything, corporations' reliance on constant production and consumption of information makes the contemporary internet exceptionally well suited to generating the conditions for digital task saturation among internet users.

THE MECHANIZATION OF SPEED IN POLITICAL LIFE

Though Hannah Arendt in *The Human Condition* does not take up technology as a central theme of the work, she does engage with technology in a way that speaks well to the later thinking of Virilio. For instance, and perhaps for literary effect as much as for analytical punch, Arendt begins the book with some marveling at the launch of Sputnik as a representation of the arrival, truly, of the age of *homo faber*, who attempts to escape the world through the manufacturing of comforting things, rather than engaging in the world as a political actor (Arendt 1958, 1–4). Read through Virilio, Arendt's comment that "men everywhere are by no means slow to catch up and adjust to scientific discoveries and technical development, but that on the contrary, have outsped them by decades," her work can be read into a dromotological analysis. *Homo faber* does not fear technology; in fact, the use of technology is key to *homo faber*'s relationship with his world. It is precisely by using "implements and tools" that *homo faber* remakes the world, working such that "everything is judged in terms of suitability and usefulness for the desired end, and nothing else" (ibid. 153). Using technology to make raw materials useful ever more quickly, *homo faber* manufacturers the conditions of his political life:

> The point is that *homo faber*, the builder of the world and the producer of things, can find his proper relationship to other people only by exchanging his products with theirs, because these products themselves are always produced in isolation. The privacy which the early modern age demanded as the supreme right of each member of society was actually the guarantee of isolation, without which no work can be produced. . . . This isolation from others is the necessary life condition for every mastership which consists in being alone with the "idea," the mental image of the thing to be. (ibid. 161–62)

This image of the worker in isolation is a curious one, considering the physical proximity necessary in most kinds of manufacturing labor, especially in the early-modern era. Arendt preempts this curiosity, pointing out that during work, communication and physical proximity enable the accomplishment of work, not "the specifically political forms of being together with others, acting in concert and speaking with each other (ibid.). A dromotological analysis of the state of labor leads to a similar conclusion. When people focus on work, there are only so many other things that they can effectively do. In worktime conversations, exchanging banalities about the weather and sporting events may well help the hours pass faster and build something like a friendship between workers, the level of conversation is substantively different from the "specifically political" conversation that Arendt holds key to a full political life. To have the time and the freedom to participate in politics, at least in the ancient world, meant having one's time freed thanks to the (enforced) labor of others (ibid. 30–31). It is necessary, in a dromotological reading of the public realm as Arendt understands it, to have not only, as she states it, the physical proximity to one another that enables political speech (ibid. 201), but also the available time to dedicate to the contemplation, listening, and responding that is necessary for action. Activity without the necessary time for reflection, judgment, and contemplation is, for Arendt, at best, mere behavior-oriented at survival rather than politics between persons who recognize one another as equal citizens (ibid., 289–94, 41–42, 45). Becoming empowered to deliberate, then, requires available time in which we can apply our intellectual faculties to deliberately political speech, not merely conversation to pass the day at work. Instead, people may well be prosuming the conditions in which they become disempowered from having the free time necessary to think and deliberate.

For Arendt, certain technologies are the core of the world of *homo faber*. Central to understanding the centrality of those tools to political life, however, is her view on both machines and tools. Before explaining that view, however, definitions are, indeed, important here considering that Arendt uses "tool" and "machine" as distinct terms. For Arendt, as we find in *The Human Condition*, a tool is something that helps a person accomplish an instance of labor or work that he already intended to do by "strengethen[ing] and multiply[ing] human strength. (ibid. 121–22). This specific formulation of the concept of the tool limits tools to a set of objects that magnify human effort in terms of speed or force. Machines, for Arendt, are different than tools.

Reading Arendt through Virilio, this is another way to examine the creation of the modern social sphere: *homo faber* lives a life at increasingly high speeds, determined first by the pace of the factory, then the car and the airplane, and now, in the twenty-first century, of the internet. The result, which Virilio refers to as "desocialization," is akin to Arendt's notion of the destruction of the capability for action; in living under conditions where we have to cope with the increased speed included by machines, whether used for commerce, war, or entertainment, the time that we possess to make use of our cognitive abilities to reflect, think, and deliberate is drastically diminished. With continuous disuse of our reflective, thinking, and deliberative abilities as individuals and in groups, those abilities may well degrade over time.

The contemporary internet presents a similar pattern. With the feudalized internet that I described in the previous chapter, people become users using digital platforms. Users, further, relate to those platforms as tools beyond the second watershed: people produce and consume the data that generate profit for those corporate actors to expand their potential digital power and profits. This casts the user in a disempowered light vis-a-vis the corporate actors who own the platforms; adding in the effects of task saturation on the users' ability to critically evaluate information critically and the result, I argue, is one in which the user has little cognitive resources or time left available with which to participate in online life to ends other than production and consumption. Technologies that are not convivial to the ends of people are then deployed directly against the people who work at those machines to empower them only as producers and consumers and disempower them from using those technologies to relate to one another through deliberation.

Here I again turn to Virilio's dromotological thinking to gain leverage on the importance of the temporal relationships people develop with technology. Certain uses of technologies foreclose possibilities for people to do activities other than production and consumption. The dromotological problem is two-fold in a condition of technologies that are beyond the second watershed. First is that the use of these technologies, by the virtue of being past the second watershed, demands that their users act following ends that are not their own; this robs the user of the time that Arendt finds so necessary for contemplation, speaking, and, ultimately, political action to be possible. Second, in the case of the internet, there is a problem with too much information. Finite human beings have only so much time and so many cognitive resources with which to think at all; far less when there is a nearly infinite source of information at our fingertips. So, not only do we have less of our own time to think and deliberate, but there is also much more information to consume in even less time.

I do not suggest that if people were free from a feudalized internet they would, or even should, spend their time contemplating politics and acting in an Arendtian sense. Instead, I offer here that users operating under task saturation on a feudalized internet have their capacity to contemplate politics or participate in democratic political life reduced by their participation in a feudalized internet that favors corporate actors. One can imagine that, after the day's tasks of formal production and consumption under a capitalist mode of life, a person could choose to garden, go for a walk, or sit to play a board game with friends. Any of these activities could, in Arendtian terms, come at the expense of time devoted to political contemplation and action. But though hobbies may detract from political time, they do not infringe on the quality of politics in the same way that participation in a feudalized, corporate internet does. Where gardening, walking, or playing with friends may not have explicitly politicized content, the feudalized web does. Gardening does not masquerade as political participation, whereas sharing a fake news story does. This masquerade is more difficult to see through when the sharer and reader of fake news may well share and read the piece under limited, task-saturated cognitive resources, and, thus, will go on to negatively affect their abilities to think critically and participate in politics as thoughtful democratic citizens.

This diminishing of available time and the disempowerment from the demanding cognitive activities that deliberative democracy demands are psychological phenomena, which I have thus far referred to through temporal compression and task saturation. I hypothesize that as human beings have more to do and less time in which to do it, our ability to do so deteriorates. That condition is not only applicable to the driving of cars and the flying of planes, but, I argue, applies just as well politically: conditions of temporal compression in the political realm result in a task saturation that disempowers us from being able to use our abilities of thought, reflection, and deliberation effectively.

Taking the work of Virilio into conversation with an adapted-ANT analysis, the condition of contemporary politics is concerning. Citizens, cast as prosumers in their relationship with corporations through digital platforms, spend much of their time consuming and producing information that makes, and keeps, them terrified of one another. This ever-accelerating pace of prosumerism may well lead some people to something not unlike a plane crash or train derailment; when politics occurs at the speed of light, some integral accident is almost certain to occur.

Chapter Three

Thinking Politically through Digital Task Saturation

THE POLITICAL PSYCHOLOGY OF INTEGRAL ACCIDENTS IN THE DIGITAL ACTOR-NETWORK

Psychologist and economist Daniel Kahneman spent much of his career studying how, exactly, human beings make decisions in the complex conditions of modernity. His work *Thinking, Fast and Slow* summarizes decades of empirical research on human beings' attention spans, and how people handle overwhelming feelings and tasks. In Kahneman's analysis, our minds process thoughts in two systems:

> System 1 operates automatically and quickly, with little or no effort and no sense of voluntary control. System 2 allocates attention to the effortful mental activities that demand it, including complex computations. The operations of System 2 are often associated with the subjective experience of agency, choice, and concentration. (Kahneman 2011, 29)

These two systems of thought, though they occupy the same physical brain, operate at different levels of effort on our part. System 1 thinking takes care of simple information and occurs without conscious effort. One example provided is the same one that I used to explain task saturation: driving a car in good conditions (ibid. 30). System 1 is generally capable of getting us through the mundane and routine actions of the day, but it has some serious downsides that lead to poor judgments and decision making on our part. Because system 1 thinking is done mostly on the subconscious level, it is vulnerable to "priming," where ideas can be placed in a context to sound one way or another, or "anchoring," where we make estimations on existing but irrelevant data (ibid. 163–64). This means, in effect, that we feel as though we are making decisions without prior thought; we are reacting with previous information in mind even if that information is not directly at the forefront of our conscious minds. In other words, we do not make our subconscious decisions with a blank mental slate but come loaded with preprocessed information.

The political ramifications for the use of system 1 thinking in terms of the capability to deliberate thoughtfully are clear: system 1 thinking is something like our intellectual autopilot that allows us to focus more consciously on other things. This relegation of system 1 thought to the intellectual background makes it susceptible to fault along several important lines. First, because it is simpler in its thought processes and actions, simple information is handled and retained by system 1: a slick slogan is more likely to stick in the back of your mind and be interpreted as the truth than a complex idea if you are in system 1 thinking (ibid. 87).

Aside from favoring simplicity, system 1 also makes note of information that makes the individual feel threatened, cataloging that information to react in ways that attempt to resolve that threat, real or perceived; this lends people to react to threats with attempts to control or end the perceived threat with violence or force rather than discourse. Further, since system 1 is responsible for actions that do not require much conscious thought, it is not a system that carefully judges; skepticism is a much harder process and, thus, saved for system 2 thinking, upon which I will elaborate in a moment (ibid. 111–12). Instead, system 1 quickly accepts that which it thinks to be true and discards other information. System 1 thought processes rely on heuristics; this makes system 1 less costly in terms of the resources it takes up to process information but leads it to fast judgment and imprecision. This is a useful trait when, for instance, one must take quick action to avoid an accident. While heuristics and the information that we maintain from priming and anchoring are highly useful when, for example, one needs to drive a car down the road; it is a system of thinking that relegates its subject matter to the subconscious. When that subconscious attention does not consider and think critically through information, accidents can occur. Task-saturated crashes in planes and cars are an example of system 1 not being able to cope with a complicated situation. Given Kahneman's analysis of system 1, I conclude that it is a system that would not serve people well in terms of democratic deliberation; relying on anchoring, priming, and heuristics places the difficult intellectual work of thinking, judging, and discoursing into a part of the mind that is more suited to helping you find your way around your home in the dark than it is to thoughtful deliberation or consideration of complex information.

As Kahneman summarizes the relevant research,

> To conserve this glucose for the more difficult tasks, our brains delegate as many tasks as possible to system 1, saving the energy for brief bursts of system 2 activity that takes more energy. (ibid. 60)

This is not always a bad thing; in fact, Kahneman surmises that intelligence is, at least in part, an effect of a brain that is efficient in delegating appropriate tasks to system 1, reserving system 2 for more effective use when needed (ibid. 64). The effects of system 2 running low on glucose are familiar to quite a few of us. The feeling of being tired after a lot of intellectual activity, reading or participating in a class, or, in the case of musicians, performing for a live crowd, with its attendant loss of cognitive ability and need to refuel with food and sleep demonstrate what happens on a quotidian level when system 2 runs our brains low on glucose.

These two systems operate in our brain simultaneously: system 1 manages the tasks our brains perceive as simpler and easier to handle, while system 2 tackles the problems and tasks that require conscious thought, finesse, or higher levels of abstraction beyond those that our heuristics and background knowledge provide. But, speaking to actor-network theory and its insistence that materiality plays a role in a person's agency (Latour 2005, 70–74), these two systems interacting contribute to the thinking and behaviors of human beings through the management of a finite pool of focused attention, dependent on the presence of reserve glucose in the brain. If corporations, computers, and prosumers are agents within the actor-network that we now know as the internet, so, too, must be the biochemical systems that make up those actors' brains. The fact that I do not consider each and every important actor in a given work does not reduce that importance but merely reflects the temporal limitations of my writing and your reading of said work. Kahneman's findings explain the plane, automobile, and surgical accidents that I discuss in terms of temporal compression and task saturation. While people can do both simple and complex tasks simultaneously, making use of the symbiosis of systems 1 and 2, doing so burns our reserve glucose quickly, leaving our minds running on system 1 alone when system 2 cannot muster the energy necessary to deal with further complications. So, when system 2, in the case of the plane crash, cannot deal with a deviation from a predetermined checklist for landing, system 1 takes over and follows the checklist straight into the countryside at a high rate of speed. Generally, when our brains cannot adequately cope with new, complex information through system 2, we revert to the information and ideas that were previously primed and anchored into our system 1 modes of thought.

Bringing Kahneman's work more explicitly to politics, I claim that there are elements of political identity that are readily available for use in system 1 thinking: party, class, and racial identities come immediately to mind here.

When availability entrepreneurs exploit people's system 1 thinking with the aforementioned continuous flow of worrying news, they are likely to do so along the lines of thinking that people are already primed to accept through their system 1 thinking; sharing a story that expresses worry about some other identified in terms of class, race, nationality, or gender is an act that uses system 1 thinking, and further keeps the sharer thinking in system 1 since the flow of such information is continuous by design.

In terms of the ability of individuals to become empowered to deliberate with one another, the findings of Kahneman are troubling. Given that deliberation requires enough time that, as Arendt develops in her telling of the ancient world, a life of labor precludes one from a fully political life, then these availability cascades, which drain glucose necessary for sustained system 2 thinking, place another barrier to political participation. If politics, through these availability cascades, is left to system 2 thinking, then we are not only excluded from a robust digital public in terms of structure but in terms of the mental ability to engage with it.

Sunstein describes the structures that Kahneman finds concerning in modern media in a series of books, all similarly titled, culminating in 2017's *#Republic*. In it, Sunstein outlines his position that deliberative democracy depends, informationally, on two things. First, "people should be exposed to materials that they would have not chosen in advance," and second, "many or most citizens should have a wide range of common experiences" (Sunstein 2007). The main crux of his argument is that this is not actually the case, and instead the web, because it is so customizable, becomes tailored for each user's preferences through a process of filtering.

In groups, this filtering results in what Sunstein refers to as "cybercascades," which occur when groups of people online come together to share similar ideas, and in so doing often convince themselves and each other that their viewpoints are simply true while leading one another to increasingly extreme versions of the same views they already had. People do not, in his line of thinking, do this consciously; we filter because our resources of attention are scarce (ibid. 25). This is in keeping with Kahneman's findings, that we can only do so much difficult or specific thinking. Rather than system 2 thinking, we make use of heuristics, such as our existing preferences to decide where to go and what to look for online. In Sunstein's view, people engaging in this behavior aid in the creation of cybercascades. Sunstein wavers in his consideration of cybercascades as a threat to democracy; at some points, he refers to their structure as a "Balkanization" of the internet, and he even devotes an entire chapter to the discussion of cybercascades as a powerful recruiting tool for terror groups. In light of the gravity of cybercascades being used as a means by which new terrorists can be recruited, it is puzzling that Sunstein's solutions remain relatively hands-off: instituting the fairness doctrine from the television broadcast industry's past, increasing funding for government websites, and hoping companies will self-regulate their platforms (ibid.). This leaves us to deal with the reality of real-time political life in which people encounter one another online in a condition of task saturation, running on system 1 thinking, and ill-equipped to sort out truth from fiction in corporate platforms that encourage users' feudalized labor.

In real-time politics, we remain, at least structurally, in much the same position as the voyeur-voyager watching the horizon in their automobile's windscreen, but because the speed of travel is the speed of light, the temporal compression and task saturation have further altered our relationship with the physical spaces around us. Instead of, for example, having several seconds to contemplate an object half a kilometer down the road in front of us, making our relationship with our current position fleeting (imagine, for example, trying to notice a single flower on the side of the road while driving at highway speeds), information confronts people at light speed. This is what Virilio terms the "anachronistic instant" in *The Great Accelerator*. Because we are moving so fast at any one given instant, notions of past, present, and future become dilated and we, as individuals bound by human bodies and used to biological conceptions of time, lose our orientation both physically and temporally. This physical disorientation, which Rosenau develops in *Distant Proximities,* is enough to cause some to retrench into tribalism as a reaction to the fear that they feel from the perception of the world is much larger than it was pre-internet; add to that the dromotological effects, and we have created in digital politics a person who may lack the skills or the perspective with which to interrogate the claims of others for validity, relying wholly on a heuristic device to make some attempt to interpret the horizons in front of them. When faced with a temporal horizon that we cannot adequately comprehend, some people react in a way that is like task saturation or system 1 thinking, and some of their heuristic devices are sexism, racism, and, in some cases, political violence.

Sunstein's understanding of the structure of the internet poses that most people encounter the internet as consumers of goods, services, and information. These consumers find their way into cybercascades through their choices as such; a consumer chooses to read articles, blogs, and so on that they like. Those articles and blogs typically feature links to other, similar pieces that the reader might find worth a click. This is an important component of how people participate in the digital world, but Sunstein's reliance on the consumer aspect of digital participation ignores a vital aspect of a feudalized internet, that these same consumers are also the producers of the content and contexts that they consume and inhabit. The idea that a person can be simultaneously a producer and a consumer is not new—Alvin Toffler coined the term *prosumer* to describe exactly this in 1980 (Ahulwalia and Miller

2014). Over the decades since then, prosumers engaged in a widening set of activities that, with the emergence of digital prosumption, create a relationship between prosumers and corporations that leave the former increasingly unlikely to have the time or mental energy to devote to the careful listening, thinking, and responding that is so central to democratic deliberation.

Pre-digital prosumption focused on the ability of a consumer to choose or customize products to fit their wants and needs while simultaneously cutting the labor costs on behalf of the producer; choosing one's produce from a supermarket shelf was a prosumptive innovation over older models of the general store, where a clerk would provide the customer with goods at the counter (ibid.). Similarly, allowing drivers to pump their gasoline was billed to both save drivers time and the stations money on staff, the savings of which could be passed down to customers. Now, prosumption has become a central element of life under capitalism—we pick our food at farms, bus our tables at restaurants, and happily assemble furniture with frustratingly minimalist pictorial instructions. These non-digital acts of prosumption share a common theme: the prosumer gets a physical product that they have chosen to fit their desires. From this perspective, early proponents of prosumerism such as Toffler saw it as a mechanism by which people could democratize economic participation by having a meaningful say in the making of the products that they would use in daily life, while producers would see reduced costs in terms of labor (Banks 1998).

People who come to believe fake news typically encounter it when someone in their digital social sphere shares it. Without prosumers to platform fake news by sharing it, social media platforms lose the content and interactivity that make them different from a simple blog. Reading some of Sunstein's conceptions of the cybercascade with the prosumer at the center of the analysis, a clearer image of the spread and persistence of fake news emerges. A cybercascade populated by prosumers is an environment that is in constant creation and reinforcement by its members. More than a group of like-minded consumers who understand one another as people who are united by shared tastes (as one would find, for instance, in a line at a salad bar), digital prosumers are the co-creators of a shared digital environment. Some of these prosumers are prodigious in their efforts. As Andrew Guess, Jonathan Nagler, and Joshua Tucker found in their study of Facebook users, just under 10 percent of accounts share fake news, but those 10 percent are responsible for sharing more than 75 percent of fake news links (Guess, Nagler, and Tucker 2019). In the context of pre-digital consumerism, this is cause for concern in terms of potentially misleading people. Claims that a household cleaner, if injected, is a replacement for vaccines, if repeated often enough and from a well-liked member of a community might cause someone to make poor health choices, such as ingesting household cleaners in an attempt to treat a respiratory virus (Kluger 2020).

Within the more directly political context of a cybercascade made up of prosumers, this sharing of fake news is more troubling. When people participate in a cybercascade, the product of that participation in the cybercascade is more links and other content that keep those people, and others, engaged within that cybercascade; this is core to the business model of social media platforms, which depend on their users to not only consume content including ads but to make much of that content for each other to see. Thus, fake news and other forms of information give the cybercascade its character and meaning for those who exist within it. In acting as a prosumer in a cybercascade, people create, re-create, and reinforce the ideas that they like, share, and follow. Much of the fake news shared thus far in the American context centers around political figures (Tsfati et al. 2020). In sharing political content, truthful or not, people make cybercascades into politically relevant spaces in their prosumption of that cybercascade. The thing prosumed, then, is a political space where members come together to share the love, or loathing, of people and events that they hold in common. Vitally, new empirical work finds that people share fake news with which they generally agree, and do not take much time to fact-check it (Geeng et al. 2020).

In light of the temporal and psychological thinking, I have covered thus far in this chapter, I argue that this sharing of fake news with which people agree is because the sharers rely on heuristics such as liking or trusting a source of information, allowing their system 1 thinking to take over without a need for critical thought. In the context of cybercascades and other epistemic bubbles, this also insulates people from a difference in terms of background and points of view that would require some more careful and resource-intensive thought to evaluate critically. Prosumers do not share news to debate its facticity or to deepen discussion; they share news, in effect, to reinforce an individual's membership in the community of people with whom they agree and have found a common space online. Engaging in communities and platforms in which one is already a member takes less time and effort than seeking out new communities, especially ones with which one disagrees. Users who are task saturated and operating within mostly system 1 thinking are likely prosumers of fake news even if they do not share lies intentionally. They so do because it is an action made easier by the programs of action built into corporate internet platforms, and within those programs of action, users run short on the cognitive resources necessary to critically evaluate information before prosuming it.

This is, vitally, an "informed and conscious" process (ibid. 32); it is there where the cybercascades of the feudalized web shift to disempowerment. Prosumption in social media platforms, especially considering Sunstein's formulation of cybercascades, may well be disempowering people from democratic participation because people encounter those cybercascades, not on an informed and conscious level more in line with system 2 thinking, but less consciously, using their system 1 thinking. Sunstein, in an article that aims at explaining social cascades, published before his work on cybercascades, points out an important phenomenon: people are willing to accept information that is not factual or change their minds to fit perceived group preferences, even when they know and can articulate privately in interviews that they know the information or opinions are not the ones that they believe (Sunstein 2000). People, Sunstein asserts, will likely conform to the prevailing views in a group, in part to retain feelings of membership and inclusion in that group, regardless of the validity or factual bases for those views (ibid.).

This mechanism is unsurprising given Kahneman's work. Recall that most of the time, people operate with system 1 thinking for easier tasks, reserving the more demanding system 2 thinking for complicated tasks. Thus, we make most of our decisions nearly instantly, operating on prior information, and heuristics such as shared social norms and ideological agreement. If then, someone who a person already likes shares a fake article about a candidate that both people abhor, the second person is more likely to simply assume it is true and spread it. This is quick, easy, and generates a feeling of having participated in a community that, as Sunstein notes in his work, is a key element of people remaining in, and believing the information presented within, cascades both online and off. It is also a mechanism by which people are prosuming a cybercascade that is directly political and disempowers its members from engaging in critical debates around the meaning of facts. Instead of seeing news, fake or real, as a set of facts about which debate can occur, the fact that people prosume news they like generates an emotional attachment to the story and sentiment shared in the fake news. Because people are doing the prosuming, which takes time, they become emotionally invested in the narratives of fake news if, for no other reason, it is easier for us to continue to do the things we already do than it is to change. This prosumption, which, I find, is a program of action baked into online platforms by corporations, conditions several categories of activity undertaken by users. Some users, such as those who spend sometimes years "researching" conspiracy theories work remarkably hard through countless hours of prosumption. The lazier prosumers, then, may simply uncritically read and sometimes share the falsehoods prosumed by their more industrious counterparts.

Because people are laboring in, and not simply consuming their cybercascades, it is even more plausible that abandoning cybercascades, and fact-checking the content there, will get less likely over time and with more participation. There is some empirical evidence that suggests this to be the case. Even after a multitude of challenges to the 2020 election were thrown out of courts across the country after so-called audits were conducted with no evidence of systematic voter fraud, lawyers for Trump suffered clear losses in courts; around a third of Trump supporters as of June 2021 still believed a recent piece of fake news, namely that he would be, by some means, "reinstated" as president by August 2021 (Morning Consult and Politico 2021). Participating in cybercascades and fake news is easy, and the emotional attractions of both are simple to understand. A person has but to open up their smartphone, scroll to something on a social network, and tap "share." In just a few seconds, the person validates their views and spreads fake news, they feel more welcome and entrenched in a political ideology that has just had one more falsehood woven into its fabric. Prosumers make their entrenchment all the easier through the unfriend and block features on these platforms: prosumers silence opposing views from interfering with cybercascades.

The undemocratic potential of this prosumption is one of the elements of contemporary capitalism online that is most concerning to Shoshana Zuboff, who, in writing *The Age of Surveillance Capitalism*, expresses deep concern that the growing amounts of data that users prosume, but do not themselves control, are put to use by corporate actors that are willing to subvert not only deliberative democratic norms but also to influence electoral outcomes in the name of increased profit and centralization of digital power.

The ideological component of the phenomenon is that constructing and playing in a reality founded on untruth has been a staple in conservative media for decades. In the cable news era, Bill O'Reilly exemplifies this well. Through his show, O'Reilly constructed a persona and ideology focused on the evils of liberalism as long-standing performance art turned propaganda believed by viewers (Baym 2009). This same line of reasoning has deepened and evolved in digital space. For instance, far-right pundit and activist Alex Jones has argued that his show is absurd and should be viewed as performance art done by a character with whom he happens to share a name, likeness, and political views (Rhodan 2017).

A range of conservative pundits and political actors, including Tucker Carlson (Folkenfink 2000), Alex Jones (Smith 2017), and the former president's former attorney Sidney Powell (Weiss 2021), have argued, sometimes in court, that the lies they tell are entertainment, not news, and that no reasonable person would believe their claims to be fact. And, to that end they are correct. No reasonable person would believe, for instance, that Hillary Clinton ran a satanic cult in the basement of a pizza parlor, or that former President Trump would be "re-instated" in August of 2021. But a person who had been a daily consumer of dubiously factual information (in the pre-internet, cable TV era after the fairness doctrine expired in 1987 [Pickard 2018]) became a prosumer in the feudalized internet. These consumers-turned-prosumers have been exposed to three decades of shifting falsehoods that made them concerned that some big *other*—whether, immigrants, women, communists, or another—was coming to destroy their prosumed version of American culture. With several decades of such exposure, especially with limited cognitive resources being pushed beyond critically important limits, some people may begin to believe some outlandish claims. Over time, Sunstein finds in *#Republic*, these prosumptive practices of sharing agreements and silencing disagreements lead to increasingly extreme versions of political ideologies finding persistent homes online.

One major concern for the disempowerment of individuals aside from the speed at which the information comes to the prosumer, to which I will return shortly, is the total amount of time that people under late capitalism spend producing. While we do, indeed, have a great ability to customize consumer goods and cybercascades, there is, I argue, a critique of prosumerism that considers political temporality as an element of disempowerment. As John Crary develops in *24/7*, Toffler was half right in thinking that prosumerism would be highly inclusive and empowering. In broad terms, Crary updates Lenin's argument from "Imperialism, The Highest Stage of Capitalism": the 1918 version of the argument posits that capitalism will continue its geographical expansion until the whole planet carries on with production, consumption, or both. Crary updates this for the twenty-first century, making the case that the newest frontier for capitalist expansion is not only geographic but also temporal. Time spent asleep in the digital age of capitalism is a missed opportunity from the perspective of those seeking profit. Seeing sleep first as a necessary evil, Crary notes, states began serious attempts to eliminate sleep to forward the destructive ends of warfare (Crary 2014). Here, Crary speaks well to Virilio's shared focus on the industrialization of war at high speed in World War II. In that conflict, belligerents on both sides of the conflict took what we would now call methamphetamine to make long flights and infantry advances possible (Andreas 2020).

For the time being, this trend toward prosumption only appears to be strengthening, with corporate actors vying to colonize even more of our time and space. Take, for example, Amazon's new Sidewalk program: any Amazon user within range could use the devices to access Amazon's goods or services. It will link together all Amazon devices, including cameras that users have placed in and around their homes. Now anyone with one of the interactive speakers sold by the company prosumes one of the largest companies in the world, effectively making many homes into storefronts. This increasing trend toward prosumption disempowers people from participating actively in democratic politics and leaves them vulnerable to fake news because they are operating under task-saturated thinking that makes it harder to critically evaluate information. In light of the psychological discussion earlier in the chapter, people who have more information, less time, and less sleep with which to muster cognitive resources to critically think, are likely to fall back into using heuristic devices like party identification rather than fact-checking the news that they like and share. Keeping in mind the Habermasian ideal that I outline in chapter 1, as well as the incidents of corporate blocking and censoring of platforms, the encroaching of corporate-controlled devices into user's homes allows those corporate actors to influence and constrain behavior in ways that are instrumental, not communicative as we would prefer in an empowered democratic public. Bringing Kahneman into conversation with Crary, the expansion of production and consumption into increasing amounts of people's time must drain the cognitive reserves that complex, system 2 thinking that democratic politics requires of its participants. With those resources stretched thin and our time for reflection, rest, and thought compressed into fleeting moments, people are likely to slip into system 1 thinking, which is more likely to entrench them into prosumed cybercascades that sometimes contain and spread fake news.

In the larger actor network of the internet, then, the prosumers as we see them today exist on the network in a state that disempowers them from making sense of the information with which they are presented. When the corporate actors, with no obligation to tell the truth as they have admitted in legal proceedings, flood cybercascades with untruths that are then prosumed at near light speed, it is little surprise that these prosumers are scared and sometimes angry. That fear and anger have dangerous real-world implications.

THE JANUARY 6 UPRISING AS AN INTEGRAL ACCIDENT OF TASK-SATURATED POLITICS

When people who are already set in their ways and feel overwhelmed, anxious, and poorly equipped to lean on the critical faculties of system 2 thinking encounter lies and untruths that confirm their preexisting views and beliefs, they react in the ways that Virilio feared, in an instantaneous politics of reactivity that leaves little time for thinking and reflection at all. The results of cybercascades, prosumed by people operating under conditions of task saturation and temporal compression, using tools that empower corporate actors to increase their access to the daily lives of people came to the public fore on January 6, 2021, when a mass of several thousand Trump supporters stormed the Capitol of the United States of America, signaling their intent to murder elected officials in an attempt to stop the count of Electoral College votes that would cement their chosen candidate's loss in the recent election (Associated Press 2021).

Composed, rational debate did occur after the election. There have been lawsuits and recounts, all of which have, through legal and civil means, reaffirmed that Donald Trump lost the 2020 election, which Trump himself has since admitted publicly (Solender 2021). What we saw on January 6, however, was exactly what Virilio had feared: politics at the speed of light as a state of thinking panic, and reaction, spread by machines of our own making, which pitted finite human capabilities against lies spread at nearly infinite speeds from the cybercascades that the people who stormed the Capitol had themselves helped to create. Virilio again gives some insight into the processes that occurred in this feudalized actor network of cybercascades leading up to the attempted overthrow of the American government on January 6, 2021.

Speaking to the consequences of political life lived at near-light speeds, Virilio finds "emotional synchronization" and a "communism of affects." These people were dissuaded from the use of their fuller critical thinking capabilities by the prosumptive programs of action built into feudalized online platforms, engaged in political violence. It was a moment, I argue, that cannot help but be one of the integral accidents of politics in the digital age; with politics reduced to a politics of the instant that people experience in "nanochronologies (Virilio 2006, 71–72) people abandoned careful thought for political reflex taught to them, the insurrectionists, for years beforehand.

The fact that the insurrectionists committed their acts without reasonable thought, and instead acted on reflexes based on absurd lies, is not merely speculation of mine or Virilio's. Some of those who fomented the insurrection and its antecedent ideology, including Alex Jones (Borchers 2017) and Sidney Powell (Polantz 2021), have argued, in courts of law, that no reasonable person would believe the clear lies that they tell.

Again, this is correct: no reasonable person *would* believe that the Democratic Party, whose leadership is supposedly running a sex cult out of a pizza parlor, managed to rig an election by counting all the votes. Some of the key ideological figures, such as the CEO of a pillow company, claim that: "We [The United States] were attacked by China, and they flipped this election and down-tickets to the tune of tens of millions [of votes]" (Wade 2021). One of the aforementioned performance artists who happened to serve as one of Trump's lawyers claimed that several agencies of the federal government had cooperated with universities to develop secret voting servers that weighed votes against the then head of that federal government less than the votes for Joe Biden (Vilareal 2021).

All these claims are outlandish. None of these claims have borne out, whether through lawsuits, congressional investigations, or publication of empirical evidence to suggest that China, Venezuela, the federal government, universities, or any other group aside from the voting public of the United States of America affected the outcome of the 2020 presidential election.

How, then, could thousands of people have been so thoroughly convinced of such outlandish claims that they would take the time to travel to Washington, DC, to challenge the election and attempt to kill sitting members of the government?

The January 6 insurrectionists, I argue, are the result of a long process by which people, participating under conditions of task saturation on a feudalized internet that encourages their continuous participation to generate more profits and data, incorporate increasingly outlandish political beliefs as part of their political heuristics.

That Trump won the election was not a matter of fact; it became, through incorporation over time into the insurrectionists' system 1 thinking, a part of their political identities, which they chose to defend at all costs. The insurrections accepted "facts" such as the lies mentioned in the previous paragraph as a matter of course; to the task-saturated mind, new information that fits old patterns and heuristics fits into existing frames of mind without burning valuable cognitive resources. For some internet users, Christopher Bail and others find in a study of social media's effects on political polarization, exposure to information proposes opposing views to that of an individual user is not seen, as Habermas would hope, as an opportunity for deliberation but as a chance to become more extreme in one's own ideology having rejected the opposing view entirely (Bail et al. 2018). After several years of incorporating falsehoods and lies as key elements of their political identities, the January 6 insurrection was, I argue, the result of minds altered over time to believe in and hold dear a political unreality that, in the eyes of those insurrectionists, was under grave and existential threat. The insurrectionists thought, in this moment of unreality, that they had to meet these threats with force if their way of life were to continue. The defense, that these users have been disabused of their conscious powers, has become a popular one among those now facing the consequences of their political reflexes (United States Attorney's Office District of Columbia n.d.).

These people acted on reflex because they were operating, for years, under a situation of task saturation and compressed temporality such that their current view of themselves and the world was a constant state of siege by perceived enemies: entire historical narratives could be, and were, crafted whole cloth by corporations and ideologies and sold to them through "news" sites and digital platforms. As Bail further develops in his illuminating work *Breaking the Social Media Prism*, the internet users that become political extremists spend much of their time (and, I add cognitive resources), feeding and participating in cybercascades that distort their perception of reality to the point that they feel as though the world around them is in need of radical change (Bail 2021). The January 6 insurrectionists chose, in their task-saturated state, to take part in the movement of a political machine that had been moving at the speed of light, and who were holding onto an ahistorical instant for dear life. It is when one views their political lives under task saturation and temporal compression that their actions can begin to make some perverse form of sense. These were people disempowered from thought and empowered as not only users of social media platforms as prosumers, but also, in that moment, combatants willing to kill and face the potential for death for their chosen leaders, who had been telling these users for years that their culture would face erasure (Pengelly 2021) by those who were committing the grand sin of counting votes in an election. These were reactive people fighting against what they saw, through the fogged screen of task saturation, as a temporal erasure of everything they held dear.

I conclude here with a moment in which Virilio speaks almost prophetically to the events of January 6 and the lead-up to them, which is an answer he gave when asked about the possibility of digital democracy:

> The tyranny of real time is not very different from classical tyranny, because it tends to destroy the reflection of the citizen in favor of a reflex action. Democracy is based on solidarity, not solitude, and man has to reflect before acting. Now, real time and the world present demand a reflex from the television viewer that is really a kind of manipulation. The tyranny of real time is tantamount to a subjugation of the television viewer. The temporality of democracy is threatened because the expectation of a judgment tends to be eliminated. . . . Live democracy, or automatic democracy, eliminates this reflection and replaces it with a reflex. (Virilio 1999, 19–20)

This is exactly what we witnessed on January 6, 2021. The integral accident of digital politics has been the replacement of reflection and thought with reflexes. The insurrectionists pulled from the ranks of people who were acting following corporate political punditry that took the myopia of the contemporary moment and weaponized it for profit. As a result, prosumers turned into insurrectionists and became so task-saturated by the compressed temporality of politics at light speed that it has become increasingly difficult for them to think rather than to react along lines learned and accepted by their system 1 thought processes, aided by cybercascades rife with untruths. Thus, the compressed temporality of digital politics has become a means by which some people have become increasingly disempowered from using their powers of reason, turning them into reactionary subjects working toward the ends of others. In terms of power in the digital landscape, this set of phenomena, I claim, tilts the balance of power toward those who own the corporate platforms on the feudalized internet who made a profit from ad revenues while these people were acting as prosumers of digital cybercascades, and from the viewership engendered by the world watching with rapt attention as the American Capitol came under siege for the first time in two centuries.

Conclusion

The Future of The Internet: New Actors and an Invitation for Empowering Activism

This book pursues two major objectives. Firstly, it seeks to make the case for an adapted-ANT analysis as a useful tool for understanding shifts in digital power dynamics. In making that case, I spend most of the book detailing the relationships between what I deem to be the major actors in the shaping of the internet so far, namely the state, corporations, and ordinary internet users. Secondly, the preceding chapters consider the asymmetries in power with the digital actor network. While the state held much of the power on the early versions of the internet, corporate power waxed in the most recent three decades. As of the second decade of the twenty-first century, corporations feudalized much of the internet; this feudalization leaves many users as prosumers, or, less euphemistically, digital serfs who labor to the ends of ever-increasing corporate power and profits. In this process of feudalization, some users' agency is so affected by their time spent in corporate-controlled digital space that they see a version of political life so immediately dangerous to their way of life that they feel it necessary to commit acts of violence up to and including an attempted coup by the supporters of Donald Trump in January of 2021.

The analysis this book presents thus far illustrates the flexibility of an adapted-ANT analysis: considering both human and nonhuman actors as co-constitutive agents in networks demand a relational understanding of each actor to develop more rigorous mappings of patterns of behavior in and between actors. Adding the critical innovation of new materialist scholarship and the historical materialist tradition from which it draws many influences adds the dimension of asymmetries of power to the analysis. Thus far, the actor network that we now know as the internet has undergone several major structural changes. What was once primarily a cooperative effort between the United States government and its contractors to reify state power along lines not dissimilar to the then-contemporary highway projects became, in the early 1990s, increasingly controlled by corporations. In the period since the commercialization of the internet in the 1990s, the states' role has diminished substantially to the point that I am willing to make the claim here that the contemporary internet has strongly feudal elements, with corporations serving as feudal lords over much of the web to the ends of generating profit. That profit comes increasingly from the prosumed labor of ordinary users. Sadly, these users-turned-prosumers are faced with a serious conundrum: the same digital spaces that are sites of prosumption are also sites in which nearly infinite amounts of information result in a state of task saturation. Such a state, the work of Paul Virilio helps explain, is one in which people are scared, angry, and prone to acting in ways that they otherwise would not. One such outcome, what I find to be an integral accident of a feudalized internet, is that some people react to their task saturation with violence.

Though January 6, 2021, was the most recent incident of mass political violence driven by web-based falsehoods peddled in no small part by corporate actors willing to make a profit off of misguided fears, I see no structural, normative, or regulatory shift that would lead me to think that the actor-network has changed in such a way that further violence on a similar, or larger, scale will not occur in the future. In the past, major change in this actor network was brought about by deliberate, large-scale investment on the part of the state, deliberate economic and normative strategies deployed by corporations in cooperation with one another, or by the global adoption of these technologies through consumer purchases of computers, phones, and tablets, and a willingness to engage as prosumers on a massive scale.

The critical history of the internet I draw thus far is fundamentally a story of change. Early internets such as ARPANET had little direct impact on the lives of most ordinary civilians; the contemporary internet, on the other hand, counts several billion of us as daily users. New actors sometimes enter into networks as well. Thus, no analysis based on either ANT or new materialism analysis can claim to be more than a mapping of some portion of an actor network from the perspective of its author. What I offer here, then, is a mapping of some of the power dynamics that shaped the contemporary internet. There are more relevant actors that I do not consider at length here. For instance, the physical infrastructure that makes up the internet or the pandemic that forced so many of us to live for over a year as primarily digital scholars, community members, and colleagues surely play major roles in shaping the current state of digital power. But as is the case with any cartographer, I must leave some details off of the map for the sake of finishing a tool that is useful to the reader, even if it is not a perfect representation of empirical reality.

Future work would do well to include more actors in considerations of digital power. In the opening months of 2023, for example, artificial intelligence

(AI) has come to the fore of corporate and news discussions (if there are meaningful distinctions between corporations and news remaining!). In the same vein that Web 2.0 was something of a deliberate marketing strategy as much, if not more, than it was an actual development and deployment of new technologies, I would urge readers to study AI critically. It is surely an introduction of new actors to the network that will necessarily change digital power dynamics, but it may be no more empowering to democratic ends that Web 2.0 has proven to be. While I am unwilling to make specific predictions of power relations on the web beyond the mapping here that holds that, in its current form, the web is a far cry from fulfilling a Habermasian ideal for free speech, I do expect that more change will come as new actors are introduced to the network.

I hope that my mapping of the contingencies of digital power is some antidote to the deterministic effect of experiencing the internet as a prosumer. Because so many of us now no longer remember a world pre–social media, let alone pre-internet, it may appear as though this is the way that things have always been and that this feudalized structure of the internet is written into our digital future. It is not so: the internet as it is now could not, I argue, have existed without decades of state-sponsored spending and research. It was a matter of legal and economic contingency within existing postwar capitalism that the current internet grew and prospered. There is a graveyard of new paradigms that corporate actors thought would, once again, reshape the digital world: virtual reality, the so-called "internet of things," cryptocurrencies, and the like. Especially when there are moments of relative corporate waning, such as during depressions or in periods where either the people organize and/or the state further regulates, there are opportunities for prosumers and users to have more say in terms of our collective digital lives.

Thus, I invite the reader to participate in this analysis; the project of mapping the power dynamics of the digital age will, I hope, consist of several snapshots such as the one that I posit in this book. With mappings drawn from multiple scholars, over time, we may begin to be able to better explain the patterns of geographical change in digital power. Once those geologies are more thoughtfully understood not only within the relatively small circles of those immediately concerned with political theory but also by those who participate in the construction of digital platforms—software engineers and prosumers—then we might begin to look again to the emancipatory potential of digital communication technologies to ends that might well allow for us to make the world a place in which we could, finally, get closer to the Habermasian ideals that undergirded much of the optimism behind early thinking about digital democracy. I am not, here, willing to predict that such a thing will occur, but the pessimism of my intellect is tempered by the optimism of my will. I have, in the course of writing this project, seen people use digital technologies to protest authoritarian governments, find and make community through a global pandemic, and stand up for their rights against those who would use misinformation and fear to further erode democracy in the United States and elsewhere. The road to a more empowering digital landscape will be a long one, but with good maps and colleagues, we might well have a chance at beginning to slow, if not reverse, the processes of feudalization that I document here. I hope that many of you will take up that journey in solidarity with one another.

Bibliography

ACLU. 2017. "Reno V. ALCU Challenge to Censorship Provisions in the Communications Decency Act." *ACLU*, June 20. Accessed October 28, 2021. https://www.aclu.org/cases/reno-v-aclu-challenge-censorship-provisions-communications-decency-act.

———. 2021. "Transcript of Trump's Speech at Rally Before US Capitol Riot." *US News and World Report.* January 13. Accessed June 28, 2021. https://www.usnews.com/news/politics/articles/2021-01-13/transcript-of-trumps-speech-at-rally-before-us-capitol-riot.

Adorno, Theodor. 2020. *Minima Moralia Refelctions from Damaged Life.* London: Verso.

———. 1991. *The Culture Industry.* New York: Routledge.

Ahulwalia, Pal, and Toby Miller. 2014. "The Prosumer." *Social Identities* 20 (4–5): 259–61.

Andreas, Peter. 2020. "How Methamphetamine Became a Key Part of Nazi Military Strategy." *Time*, January 7. https://time.com/5752114/nazi-military-drugs/.

Arendt, Hannah. 1958. *The Human Condition.* Chicago: University of Chicago Press.

Associated Press. 2021. "Video Shows Capitol 'Mob Calling for the Death of the Vice President,' Plaskett Says." *PBS News Hour*, February 10. https://www.pbs.org/newshour/politics/watch-video-shows-capitol-mob-calling-for-the-death-of-the-vice-president-plaskett-says.

Aufderheide, Patricia. 2002. "Competition and Commons: The Public Interest In and After the AOL-Time-Warner Merger." *Journal of Broadcasting and Electronic Media* 46: 515–31.

Bail, Christopher. 2021. *Breaking the Social Media Prism.* Princeton: Princeton University Press.

Bail, Christopher, Lisa Argyle, Taylor Brown, John Bumpus, Haohan Chen, MB Fallin Hunzaker, Jaemin Lee, Marcus Mann, Freidolin Merhout, and Alexander Volfovsky. 2018. "Exposure to Opposing Views on Social Media Can Increase Political Polarization." *Proceedings of the National Academy of Sciences* 9216–9221.

Banks, Eric. 1998. "The Prosumer and the Productivity Paradox." *Social Policy* 28 (4): 10–14.

Baym, Geoffrey. 2009. "Stephen Colbert's Parody of the Postmodern." In *Satire TV*, by Jonathan Gray, Jeffery Jones, and Ethan Thompson, 126–30. New York: NYU Press.

Bement, Austin. 1916. "The Economic and Strategic Value of the Lincoln Highway as Considered from the Standpoint of National Defense." *The Scientific Monthly* 373–80.

Bennett, Jane. 2010. *Vibrant Matter*. Durham: Duke University Press.

Borchers, Callum. 2017. "Alex Jones Should Not Be Taken Seriously, According to Alex Jones's Lawyers," *The Washington Post*, April 17. https://www.washingtonpost.com/news/the-fix/wp/2017/04/17/trump-called-alex-jones-amazing-joness-own-lawyer-calls-him-a-performance-artist/.

Boyd, Danah, and Kate Crawford. 2012. "Critical Questions for Big Data." *Information, Communication, and Society* 15, pp. 662–79.

Braidotti, Rossi. 2012. "Interview with Braidotti Rossi." In *New Materialism: Interviews and Cartographies*, by Rick Dolphhijn and Iris van der Tuin, 33. Ann Arbor: Open Humanities Press.

Brown, Wendy. 2019. *In the Ruins of Neoliberalism*. New York: Columbia University Press.

Bunn, Austin. 1998. "Free for All." *The Village Voice*, January 20. https://www.villagevoice.com/1998/01/20/free-for-all/.

Castree, Noel. 2002. "False Antithesis? Marxism, Nature, and Actor-Networks." *Antipode* 34, no. 1 (January 2002): 111–46.

The Centers for Disease Control and Prevention. 2021. *Distracted Driving*. Technical Report, Washington DC: The Centers for Disease Control and Prevention. https://www.cdc.gov/transportationsafety/distracted_driving/index.html#:~:text=In%20the%20U.S.%20in%202018,crashes%20involving%20a%20distracted%20driver.&text=About%201%20in%205%20of,or%20otherwise%20outside%20a%20vehicle.

Chun, Wendy Hui Kyong. 2011. *Programmed Visions*. Cambridge: The MIT Press.

Cochran, James, Ali Darrat, and Khaled Elkhal. 2006. "On The Bankruptcy of Internet Companies: An Empirical Inquiry." *Journal of Business Research* 59, nos. 10–11 (October 2006): 1193–2000.

Connolly, William. 2013. "The 'New Materialism' and the Fragility of Things." *Millennium – Journal of International Studies* 41, no. 3: 400.

Crain, Matthew. 2014. "Financial Markets and Online Investing: Reevaluating the Dotcom Investment Bubble." *Information, Communication, and Society* 17, no. 3: 371–84.

Curlew, Abigail. 2019. "Doxxing, Vigilantes, and Transmisogyny." Medium, May 3. https://medium.com/@digitaljusticelab/doxxing-vigilantes-and-transmisogyny-c2b8a6abb2b2.

Crary, Jonathan. 2014. *24/7*. New York: Verso Press.

Davis, Bradley, Katherine Welch, and Timmothy Pritts. 2013. *Enhanced Critical Care Air Transport Team Training for Mitigation of Task Saturation*. Final Technical Report, Cincinnati: USAF School of Aerospace Medicine.

Davis, Andrew. 2018. "At One Point, Amazon Lost More than 90% of Its Value. But Long-Term Investors Still Got Rich." CNBC, December 18. https://www.cnbc.com/2018/12/18/dotcom-bubble-amazon-stock-lost-more-than-90percent-long-term-investors-still-got-rich.html.

DeLanda, Manuel. 2012. "Interview with Manuel DeLanda." In *New Materialism: Interviews and Cartographies*, by Rick Dolphhijn and Iris van der Tuin, 39. Ann Arbor: Open Humanities.

Elias, Jennifer, and Magdalena Petrova. 2019. "Google's Rocky Path to Email Domination." CNBC, October 26. https://www.cnbc.com/2019/10/26/gmail-dominates-consumer-email-with-1point5-billion-users.html#:~:text=Google's%20Gmail%20is%20the%20dominant,1.5%20billion%20global%20active%20users.

Fishkin, Rand. 2018. "New Jumpshot 2018 Data: Where Searches Happen on the Web (Google, Amazon, Facebook, and Beyond)." SparkToro, April 4. Accessed January 26, 2021. https://sparktoro.com/blog/new-jumpshot-2018-data-where-searches-happen-on-the-web-google-amazon-facebook-beyond/.

Folkenfink, David. 2020. "You Literally Can't Believe the Facts Tucker Carlson Tells You. So, Say Fox's Lawyers." NPR, September 29. https://www.npr.org/2020/09/29/917747123/you-literally-cant-believe-the-facts-tucker-carlson-tells-you-so-say-fox-s-lawye.

Gamble, Christopher, Joshua Hanan, and Thomas Nail. 2019. "What Is New Materialism?" *Angelaki, Journal of the Theoretical Humanities* 24, no. 6 (November 2019): 111–34. https://doi.org/10.1080/0969725X.2019.1684704.

Geeng, Christine, Savannah Yee, and Fransizka Roesner. 2020. "Fake News of Facebook and Twitter: Investigating How People (Don't) Investigate." *Conference on Human Factors in Computing Systems.* Honolulu: Association for Computing Machinery. Paper 655.

Geiger, R. Stuart. 2009. "Does Habermas Understand the Internet? The Algorithmic Construction of the Blog/Public Sphere." *Gnovis, A Journal of Communication, Culture, and Technology* 10, no. 1 (2009). https://gnovis. https://ssrn.com/abstract=2734947.

Goodnight, Thomas, and Sandy Green. 2010. "Rhetoric, Risk, and Markets: The Dot-Com Bubble." *Quarterly Journal of Speech* 69 (2): 115–40.

Griffith, Robert. 1982. "Dwight D. Eisenhower and the Corporate Commonwealth." *The American Historical Review* 87, no. 1 (February 1982): 87–122.

Groff, James R. 2006. "Task Saturation." *Combat Edge*, December.

Guess, Andrew, Jonathan Nagler, and Joshua Tucker. 2019. "Less Than You Think: Prevalence and Predictors of Fake News Dissemination of Facebook." *Science Advances* 5, no. 1 (January 9, 2019).

Habermas, Jurgen. 1991. *The Structural Transformation of the Public Sphere*. Cambridge: The MIT Press.

Institute of Government at Florida International University. 1998. *Final Report: Historical Impacts of Transportation Projects on the Overtown Community.* Final Research Report, Miami: Florida International University.

Internet Corporation for Assigned Names and Numbers. 1991. *Network Information Services Manager(s) for NSFNET and the NREN: INTERNIC Registration Services*. Government Contract, Washington DC: ICANN. http://archive.icann.org/en/nsi/coopagmt-01jan93.htm.

Jarvis, Jeff. 2011. *Public Parts*. New York: Simon and Schuster.

Johannson, Gunnar, and Kare Rumar. 1971. "Drivers' Brake Reaction Times." *Human Factors* 13, no. 1 (1971): 23–27.

Kahneman, Daniel. 2011. *Thinking, Fast and Slow*. New York: Farrar, Strauss, and Giroux.

Kesan, Jay, and Rajiv Shah. 2001. "Fool Us Once Shame on You—Fool Us Twice Shame on Us: What We Can Learn from the Privatizations of the Internet Backbone and the Domain Name System." *Washington University Law Quarterly* 79, no. 1 (2001): 91–220. https://openscholarship.wustl.edu/cgi/viewcontent.cgi?article=1399&context=law_lawreview.

Kluger, Jeffrey. 2020. "Accidental Poisonings Increased after President Trump's Disinfectant Comments." *Time*, May 18. https://time.com/5835244/accidental-poisonings-trump/.

Koch, Nadine, and Eric H Shuckmann. 1998. "Democratizing Internet Access in the Lesbian, Gay, and Bisexual Communities." In *Cyberghetto or Cybertopia?*, edited by Bosah Ebo, 171–84. Westport: Praeger.

Latour, Bruno. 1986. *Laboratory Life*. Princeton: Princeton University Press.

———. 1992. "Where Are the Missing Masses? The Sociology of a Few Mundane Artifacts." In *Shaping Technology/Building Society*, edited by Wiebe Bijker and John Law, 225 58. Cambridge: MIT Press.

———. 2000. *Pandora's Hope*. Cambridge: Harvard University Press.

———. 2004. *Politics of Nature*. Cambridge: Harvard University Press.

———. 2005. *Reassembling the Social*. Oxford: Oxford University Press.

———. 2014. *On Some of the Affects of Capitalism. Lecture, Copenhagen*: Royal Academy, Copenhagen.

Lenin, VI. 1966. "Imperialism, the Highest Stage of Capitalism." In *Essential Works of Lenin*, edited by Henry Christman, 177–364. New York: Dover.

Marx, Karl. 1972. "Marx on the History of His Opinions." In The Marx-Engels Reader, by Karl Marx and Friedreich Engels, 3–11. New York: W. W. Norton and Company.

McGill, Andrew. 2017. "A Bot That Can Tell When It's Really Donald Trump Who's Tweeting." March 28. *The Atlantic*, March 28, 2017. Accessed September 7, 2020. https://www.theatlantic.com/politics/archive/2017/03/a-bot-that-detects-when-donald-trump-is-tweeting/521127/.

McIlwain, Charlton. 2020. *Black Software*. Oxford: Oxford University Press.

Medialive International; O'Reilly Media Inc. 2004. "Web 2.0 Conference." June 2. Accessed January 27, 2021. https://web.archive.org/web/20040602111547/http://web2con.com/.

Meltzer, Allan. 1988. "Economic Politics and Actions in the Reagan Administration." *Journal of Post-Keynesian Economics* 10, no. 4 (Summer 1988): 528–40.

Meta Platforms, Inc. 2022. "Meta Reports Fourth Quarter and Full Year 2021 Results." Corporate Press Release, Menlo Park, California: Meta Platforms, Inc. https://investor.fb.com/investor-news/press-release-details/2022/Meta-Reports-Fourth-Quarter-and-Full-Year-2021-Results/default.aspx.

Miles, Sam. 2018. "Still Getting It On Online: Thirty Years of Queer Male Spaces Brokered through Digital Technologies." Geography Compass (September 6). doi:10.1111/gec3.12407.

Morning Consult and Politico. 2021. *National Tracking Poll #210636.* Political Opinion Poll, Morning Consult, and Politico. https://assets.morningconsult.com/wp-uploads/2021/06/09050746/210636_crosstabs_POLITICO_RVs_v1_LM.pdf.

Moschovitis, Christos, Hillary Poole, Tami Schuyler, and Theresa Senft. 1999. *History of the Internet: A Chronology, 1843 to the Present.* Santa Barbara: ABC-CLIO.

Muller, Stig, and Hitendra R. H. Patel. 2012. "Lessons Learned from the Aviation Industry: Surgical Checklists." In *Simulation Training in Laparoscopy and Robotic Surgery*, by H. Patel and J. Joseph, 1-6. Springer: London.

Musser, John. 2007. *Web 2.0 Principles and Best Practices*. Sebastopol: O'Reilly Media Inc.

National Transportation Safety Board. 2016. *Crash during Nonprecision Instrument Approach to Landing Execuflight Flight 1526.* Aircraft Accident Report, National Transportation Safety Board. https://www.ntsb.gov/investigations/AccidentReports/Reports/AAR1603.pdf.

Neff, Gina. 2012. *Venture Labor.* Cambridge: The MIT Press.

Nesson, Charles, and David Marglin. 1996. "The Day the Internet Met the First Amendment: Time and the Communications Decency Act." *Harvard Journal of Law and Technology* (Fall 1996): 113–36.

Olsen, Stefanie. 2004. "Mark Cuban Raises Specter of Dot-Com Redux." c|net, October 2. Accessed January 26, 2022. https://web.archive.org/web/20050316013930/http://news.com.com/Mark+Cuban+prompts+dot-com+redux/2100-1026_3-5400029.html?tag=nefd.top.

O'Reilly Media. 2010. "Web 2.0 Summit 2010: Mark Zuckerberg; A Conversation with Mark Zuckerberg." YouTube, November 19. Accessed January 28, 2022. https://web.archive.org/web/20101206063625/http://www.youtube.com/watch?v=CRUOl03nZIc&p=2737D508F656CCF8.

O'Reilly, Tim. 2005. "What Is Web 2.0." O'Reilly Network, September 30. Accessed January 27, 2021. https://www.oreilly.com/lpt/a/1.

Panagia, Davide. 2021. "On the Possibilities of a Political Theory of Algorithms." *Political Theory* 49, no. 1 (September 2020): 109–33.

Pengelly, Martin. 2021. "Lachlan Murdoch Backs Tucker Carlson in 'White Replacement' Furor." *The Guardian*, April 13. https://www.theguardian.com/media/2021/apr/13/tucker-carlson-fox-news-murdoch-anti-defamation-league.

Pickard, Victor. 2018. *The Strange Life and Death of the Fairness Doctrine.* Departmental Paper, Philadelphia: Communication Commons. https://repository.upenn.edu/entities/publication/30e1d49e-aabc-46bb-ba45-640e57526591.

Plato. 1993. *The Republic.* Translated by Robin Waterfield. New York: Oxford University Press.

Polantz, Katelyn. 2021. "Sidney Powell Argues in a New Court Filing That No Reasonable People Would Believe Her Election Fraud Claims." CNN Politics, March 23. https://www.cnn.com/2021/03/22/politics/sidney-powell-dominion-lawsuit-election-fraud/index.html.

Reading, Don. 1973. "New Deal Activity and the States, 1933 to 1939." *The Journal of Economic History* 33, no. 4 (1973): 792–810.

Rhodan, Maya. 2017. "President Trump's Favorite Conspiracy Theorist Is Just 'Playing a Character,' His Lawyer Says." *Time*, April 17. https://time.com/4743025/alex-jones-infowars-divorce-donald-trump/.

Ritter, Jay. 2014. "Google's IPO, 10 Years Later." *Forbes*, August 7. https://www.forbes.com/sites/jayritter/2014/08/07/googles-ipo-10-years-later/?sh=4209a6292e6c.

Rollins, William. 1995. "Whose Landscape? Technology, Fascism, and Environmentalism on the National Socialist Autobahn." *Annals of the Association of American Geographers* 85, no. 3 (September 1995): 454–520.

Rose, Mark, and Raymond Mohl. 2012. *Interstate.* Knoxville: The University of Tennessee Press.

Rosenau, James. 2003. *Distant Proximities.* Princeton: Princeton University Press.

Salkin, Allen. 2009. "For Him, The Web Was No Safety Net." *The New York Times*, August 28. https://www.nytimes.com/2009/08/30/fashion/30harris.html?ref=fashion&pagewanted=all.

Shinhal, John. 2005. "Netscape, the IPO That Launched an Era." Marketwatch, August 5. https://www.marketwatch.com/story/netscape-ipo-ignited-the-boom-taught-some-hard-lessons-20058518550.

Simon, Leslie David, Javier Corrales, and Donald Wolfsenberger. 2002. *Democracy and the Internet*. Washington DC: The Woodrow Wilson Center Press.

Smith, Allan. 2017. "'It's Performance Art': Lawyer for Alex Jones Says InfoWars Founder Is 'Playing a Character." *Business Insider*, April 17. https://www.businessinsider.com/lawyer-alex-jones-infowars-playing-character-acting-2017-4.

Smith, Andrew. 2012. *Totally Wired: The Rise and Fall of Josh Harris and the Great Dotcom Swindle*. New York: Black Cat.

Soderberg, Johan, and Adam Netzen. 2010. "When All That Is Theory Melts into (Hot) Air: Contrasts and Parallels Between Actor-Network Theory, Autonomist Marxism, and Open Marxism." *Ephemera* 10, no. 2 (2010): 95–118.

Solender, Andrew. 2021. "Trump Says He 'Didn't Win' the 2020 Election and Wants Biden to 'Do Well.'" *Forbes*, June 17. https://www.forbes.com/sites/andrewsolender/2021/06/17/trump-says-he-didnt-win-the-2020-election-and-wants-biden-to-do-well/?sh=40cfd6f8403c.

Sunstein, Cass. 2000. "Deliberative Trouble? Why Groups Go to Extremes." *The Yale Law Journal* 110, no. 1 (October 2000): 71–119.

———. 2007. *Republic.com 2.0*. Princeton: Princeton University Press.

Taylor, Astra. 2014. *The People's Platform*. New York: Metropolitan Books.

Teeter, Preston, and Jorgen Sandberg. 2017. "Cracking the Enigma of Asset Bubbles With Narratives." *Strategic Organization* 15: 91–99.

Thornton, Jennifer, and Sunny Marche. 2003. "Sorting through the Dot Bomb Rubble: How Did the High-Profile E-Tailers Fail?" *International Journal of Information Management* 23 (2): 121–38.

Tsfati, Yariv, G. H. Boomgararden, J. Stromback, J. Vliegenthart, A. Damstra, and E. Lingren. 2020. "Causes and Consequences of Mainstream Media Dissemination of Fake News: Literature Review and Synthesis." *Annals of the International Communication Association* 40 (2): 157–73.

United States Attorney's Office District of Columbia. n.d. *Capitol Breach Cases.* Accessed April 14, 2021. https://www.justice.gov/usao-dc/capitol-breach-cases.

United States Department of Transportation. 2001. "Minimum Penalties for Repeat Offenders for DWI or DUI." United States Department of Transportation. Accessed September 16, 2020. https://www.fhwa.dot.gov/tea21/factsheets/n_164.htm.

Vaccari, Cristian. 2013. *Digital Politics in Western Democracies*. Baltimore: Johns Hopkins University Press.

Villareal, Daniel. 2021. "Sidney Powell Claims All Votes Go to a Secret Server So People Can Manipulate Them." *Newsweek*, October 11. https://www.newsweek.com/sidney-powell-claims-all-votes-go-secret-server-so-people-can-manipulate-them-1637840.

Virilio, Paul. 1999. *Politics of the Very Worst.* New York: Semiotext(e).

———. 2006. *Speed and Politics.* Cambridge: Semiotext(e).

———. 2007. *Negative Horizon.* New York: Continuum.

———. 2008. *Pure War.* Los Angeles: Semiotext(e).

———. 2012. *The Great Accelerator*. Cambridge: Polity Press.

Wade, Peter. 2021. "This MyPillow Guy Meltdown Would Be Funny If It Weren't So Dangerous." *Rolling Stone*, August 14. https://www.rollingstone.com/politics/politics-news/mypillow-guy-meltdown-lindell-1212379/.

Waldherr, Annie, and Stephanie Geise. 2019. "Because Technology Matters: Theorizing Interdependence in Computational Communication Science with Actor-Network Theory." *International Journal of Communications* 13 (2019): 3955–75.

Walsh, Charlotte. 2011. "Drugs, the Internet and Change." *Journal of Psychoactive Drugs* 43, no. 1 (March 2011): 55–63.

Warren, Mark. 1989. "Liberal Constitutionalism as Ideology: Marx and Habermas." *Political Theory* 519.

Weiss, Debra Cassens. 2021. "Sidney Powell Lawyers Argue No Reasonable Person Would Have Accepted Her Stolen Election Claims as Fact." *American Bar Association Journal*, March 23. https://www.abajournal.com/news/article/sidney-powell-lawyers-argue-no-reasonable-person-would-have-accepted-her-stolen-election-claims-as-fact.

World Bank. 2020. "Individuals Using the Internet (% of population)." *International Telecommunication Union, World Telecommunication/ICT Development Report and database.* Accessed July 17, 2020. https://data.worldbank.org/indicator/IT.NET.USER.ZS?end=2018&most_recent_value_desc=true&start=2018&type=shaded&view=map.

Zizek, Slavoj. 1999. "The Matrix, or The Two Sides of Perversion." Inside the Matrix: International Symposium at the Center for Art and Media, Karlsruhe. Karlsruhe. Online. https://www.lacan.com/zizek-matrix.htm.

Zuboff, Shoshana. 2019. *The Age of Surveillance Capitalism.* New York: PublicAffairs.

Index

About the Author

Garrett Pierman received his PhD in political science from Florida International University in 2022. There, his research lies at the intersection of political theory and digital media studies. He has taught in the departments of politics and international relations, computer science, and for the Honors College. A first-generation student, he is dedicated to the empowering potential of education for folks from marginalized backgrounds and hopes that through sustained critique and activism, we can begin to tackle the existential problems that our species faces in the twenty-first century.